Water Treatment Process Monitoring and Evaluation

Water Treatment Process Monitoring and Evaluation

Richard P. Beverly, PE

American Water Works Association

Disclaimer

AWWA Publications Manager: Gay Porter De Nileon
Technical Editor/Project Manager: Melissa Valentine
Cover Design: Cheryl Armstrong
Production: Sheryl Tongue

Library of Congress Cataloging-in-Publication Data

Water treatment process monitoring and evaluation / by Richard P. Beverly
p. cm.
Includes bibliographic references.
ISBN 978-1-58321-858-7
1. Water treatment plants--Evaluation. 2. Plant performance--Evaluation. 3. Water--Purification--Evaluation. I. Beverly, Richard P.
TD434.W425 2011
628.1'62--dc23

2011034828

ISBN 1-58321-858-0
978-158321-858-7

American Water Works Association

6666 West Quincy Avenue
Denver, CO 80235-3098
303.794.7711
www.awwa.org

Contents

10 Computer Control Hardware 73

11 Instrument Installation and Chemical Locations 79

12 Response Time 89

13 Operations Manual/Records 95

14 Evaluation of Process Data Over Time 101

List of Figures

List of Tables

Acknowledgments

The author would like to thank the following groups for their help in preparing this book:

Technical Information:
- The Pall Corporation
- The Hach Company
- Ric Tower, Treatment Equipment Co. of Portland, Ore.

Peer Review:
- Chuck Kingston and the staff of the Joint Water Commission treatment plant at Hillsboro, Oregon.

About the Author

Richard P. (Phil) Beverly has a B.S. degree in Engineering from the Virginia Polytechnic Institute. He has professional licenses in civil, mechanical, and environmental engineering and has held a Level III Water Treatment Plant Operator certificate with the filter endorsement in the state of Oregon. He has served on the AWWA B604 Standards Council Subcommittee for Granular Activated Carbon as Filter Media. He was also on the Environmental Registry Who's Who for 1992.

Mr. Beverly has over 37 years of experience in troubleshooting filters in both water and wastewater applications. He has helped rehabilitate approximately 200 filter plants and has been retained as a special consultant to resolve problems in numerous filtration facilities for water and wastewater treatment systems across the country and overseas.

Previous books he has written for AWWA include *Filter Troubleshooting and Design Handbook*, and *Pump Selection and Troubleshooting Field Guide*.

The author can be contacted at 541.786.0907 or RBeverlys@juno.com.

CHAPTER ONE

Introduction

The intent of this handbook is to provide information and tools to assist water treatment plant operators in monitoring plant operations and evaluating plant operational changes (such as the changes in treatment efficiency due to changes in the raw water) in order to help the operators make corresponding process changes (water chemistry, etc.) to keep their plants operating properly.

To accomplish this goal, a "building block" approach is used: First the handbook methodically builds up background information that all operators should know to effectively troubleshoot treatment problems—detailing, for example, such treatment plant fundamentals as basic design principles and design issues, process flow diagrams, data monitoring and record keeping, and instrumentation and control systems (including response time). Then the handbook delves into the main topic of the evaluation of monitoring data in detail—covering, for example, turbidity, filter-to-waste cycle, filter run time, head loss, alkalinity, pH, filtration efficiency, and the like. Case histories, plentiful graphic illustrations, and a rich assortment of appendix material augment the text.

TREATMENT PLANT DEFINITION

For the purposes of this discussion, the term *treatment plant* will refer primarily to water treatment, although wastewater filtration is also related. A conventional treatment plant is assumed for discussion purposes, although any type may be substituted, including package plants, either gravity or pressure treatment units, as well as microfiltration or reverse osmosis. All the guidelines included herein can be applied to all of these types of treatment plants, including conventional treatment used for pretreatment for membrane filtration.

OPERATIONAL NEEDS

Field experience has demonstrated that resolving operational issues may not be easy in some cases. Besides training in treatment processes, operators need a knowledge of basic chemistry, experience in their own treatment plant, and the ability to analyze process changes and make the necessary changes to maintain the desired water quality. To accurately perform these tasks, the treatment plant should be provided with maximum operational flexibility and adequate instrumentation to identify the reason for any process changes that may occur.

PROPER OPERATION

The ability to provide proper operation starts at the design level. The plant designer should have a good grasp of the raw water chemistry at the particular site, as well as the treatment processes needed to achieve the desired water quality, and finally a very good grasp of the workings of all the plant equipment. These are considerable requirements. Many engineers design water treatment plants, but the ability to include operational flexibility and monitoring to make a plant "livable and workable" is another matter.

In other words, if the plant is not designed properly, it may be difficult to operate regardless of its size. One of the many admirable qualities of many operators is that they seem to find a way to make things work in spite of the way the plant is laid out and built. Still, the designers can go a long way toward making a plant more livable and easier to operate.

DESIGN KNOWLEDGE

An operator does not need have design capability or training. However, an operator would have a great advantage in having some knowledge of process design, including such things as the desired instrumentation, the desired location of such instrumentation, the proper construction of sampling points, the necessary detention time required after chemical addition, as well as other issues such as the effect on the plant from taking one or more filters off-line for backwashing. Both operators and system managers would also benefit by knowing if their plant is lacking in the necessary analytical tools so that they can take steps to obtain them.

OPERATOR NEEDS

It is often said that experience is the best teacher. However, obtaining the required experience can be difficult, unless the necessary tools are available. An example is when a new plant has been built that uses technology or equipment different from that which an experienced operator is used to. Another example is where an inexperienced operator comes to a new plant. In such cases, a detailed operational manual would be extremely beneficial but is often unavailable.

CONTENT AND GOALS OF THIS HANDBOOK

This handbook aims to cover water treatment plant fundamentals that are essential for all operators who want to identify process problems, evaluate the cause(s), and develop effective solutions. Preliminary chapters present information about design that all operators should know including the development of process flow diagrams (a hypothetical plant is used as an example). Next, various aspects of monitoring data, such as control systems, instrumentation, and record keeping, are discussed with a special chapter devoted to response time. Finally, the handbook presents a detailed discussion of the evaluation of process data over time—covering a wide range of water treatment plant process issues.

The intent of this handbook is to provide operators and system managers with the analysis tools to understand and operate a plant and to be able to identify and correct any plant deficiencies.

CHAPTER TWO

Preliminary Design

When a treatment plant is planned, some elements—for example, a preliminary design report and a pilot study—will aid those operating and troubleshooting the full-scale plant in the future. This chapter and several following chapters will describe such elements.

The first phase of a treatment plant design is to prepare a preliminary design report. The preliminary design report establishes the scope of the treatment processes to be used; determines the estimated cost of construction; determines the chemical feed requirements; and determines the methods to be used for disposal of chemical waste, as well as other issues.

In some cases, the disposal of the chemical waste from the plant governs the type of treatment processes to be used. For example, one such case involved the treatment and removal of iron and arsenic from drinking water in a remote location. On-site disposal of wastewater was also required because of the remote location. At that location, chemicals that ordinarily would have been used were prohibited by the site constraints. An alternate treatment technology was therefore required.

PILOT STUDIES

A pilot study is often included as part of the preliminary design report. When an entirely new treatment plant is to be constructed, the regulatory authorities often require that a pilot study be performed to confirm that the preliminary design is valid and will work. The project owner and/or the entity providing the project financing may also require a pilot study.

A pilot study is extremely valuable and should be conducted. A pilot study report provides the basis for design in addition to being a valuable training tool for the operators. It is also a valuable resource for future plant operations.

PLANT EXPANSION

If an expansion of an existing plant is planned, a pilot study may not be required if the same treatment processes and equipment are to be used. However, if different treatment processes are planned, a pilot study may still be required. Some regulatory authorities require pilot studies if new treatment technologies are to be used.

PRELIMINARY DESIGN WITHOUT A PILOT STUDY

If a preliminary design report is completed without the benefit of a pilot study, the designer is assuming complete responsibility for the success of the proposed treatment processes. In such a case, a comprehensive operations manual should be prepared for the benefit of the plant operators. Such an operations manual should be provided in any case. Without a pilot study for reference, the importance of the operations manual increases. Chapter 13 will discuss operations manuals, and chapter 3 will review pilot studies.

CHAPTER THREE

Pilot Study Purpose

Pilot studies provide plant operators and system managers with a valuable resource in determining how the plant should be operated. On many occasions, operators could have used a pilot study report for help in determining how to resolve treatment issues.

There are numerous benefits of conducting a pilot study including verification of the treatment process selection, verification that the treatment processes can provide the desired water quality, sizing the treatment plant processes including the unit flow rate for granular media filters or the flux rate for membrane filtration, determining the optimum chemical feed rates and sizing that equipment, as well as operator training.

PILOT FILTER STUDIES

Pilot filter studies consist of small-scale testing to determine the treatability of a water source, to determine the effectiveness of various media types, and/or to determine the chemical feed requirements. The intent is to duplicate the treatment processes to be used for the full-scale plant. Depending on the type of pilot equipment used, the flow rate for testing purposes is usually in the range of 1–50 gpm, although many pilot test flow rates are in the range of 1–5 gpm.

In most cases, a pilot study program consists of two stages; the first is bench-scale/jar testing to establish a chemical feed baseline for the second stage of continuous process testing.

PILOT STUDY GOALS

The pilot study goals must be clearly defined before beginning any actual test work. Typical goals are discussed in the following sections.

Treatability

The treatability of the source water should be determined in terms of processes needed and chemical requirements. In other words, there may be a number of different treatment processes that may work to some extent. However, the ones that are expected to be the most effective should be piloted. Then, the pilot study results should be evaluated for highest efficiency, in terms of cost, chemical use, and disposal, as well as the quantity of wastewater produced. There may be cases, as mentioned previously, where the method available for wastewater disposal may dictate the selection of the treatment process to be used.

Effectiveness

The effectiveness of various treatment processes should be verified including any potential new technologies. For example, if two different processes will work, does one produce any other benefits, other than the basic treatment requirements? If all other criteria are approximately the same, does one produce less wastewater or does one produce a higher quality water, over and above the basic requirements? In such cases, long term benefits may be such that cost may not be the best method of evaluation, unless the cost differences are great.

Chemical Feed

The chemicals to be used in treatment and the proper feed rates for each should be verified. This information is very necessary for sizing the chemical feed equipment for the full-scale plant. There may be cases where two different chemicals will work, but one may require the use of a higher feed rate. In such cases, required maintenance and the quantity of wastewater produced may be critical factors in selecting the chemicals to be used.

Early Warning

Some plants have raw water that is subject to rapid changes. In such cases, a permanent pilot unit may be desirable, if it can provide preliminary results faster than the water that passes through the full-scale plant. With such an early warning, water chemistry changes can be made rapidly before any poor quality water passes through the plant to the final instrumentation. If poor quality water is allowed to enter the plant without adequate process adjustments, it may result in production of poor quality water and a considerable loss of production time.

Operator Training

Pilot units provide a highly useful tool for operator training, either on a temporary basis during the pilot study or on a permanent basis as part of the full-scale plant equipment. An example is conducting chemical trials to determine if a different chemical can provide more efficient treatment than the chemical currently being used. Trials can be conducted without the risk of upsetting the full-scale plant if the trial chemical does not work as well.

Another potential benefit, especially for granular media filter types of pilot studies, is the ability to visually demonstrate media fluidization and classification during the backwash cycle.

PILOT TESTING PROTOCOL

Having a testing plan or protocol is critical to a successful pilot study. The pilot testing protocol must be completed before beginning any testing. It will provide an outline for the final pilot study report. One microfiltration pilot test failed due to a lack of proper direction. The pilot test operators only ran one chemical setting without including any changes. Once preliminary lab tests were run to establish the expected range of chemical settings, the pilot test was concluded successfully.

A basic outline of items in a typical pilot testing protocol, or operational plan, is included in the following section.

Protocol

A pilot testing protocol should include the following:

- **Raw Water Quality—**The raw water quality should be determined using test wells for groundwater or samples from a surface supply, depending on the source planned for the full-scale plant.
- **Treatment Processes—**Using the desired finished water quality, the treatment processes to be tested should be determined, such as conventional treatment with coagulation, greensand, or membranes, and so on. NOTE: The pilot testing may include multiple treatment options and equipment trains.
- **Flow/Flux Rates—**A range of unit flow or flux rates to be tested should be determined. Knowing the acceptable ranges for the flow/flux rates will be used in sizing the full-scale plant.
- **Chemicals—**The chemicals needed for such purposes as buffering, oxidation, coagulation, filter aid, posttreatment pH adjustment, disinfection, and so on should be determined.
- **Bench-Scale/Jar Tests—**Bench-scale/jar tests should be performed to establish ranges of chemicals for further testing. The lack of these preliminary tests may make pilot testing more difficult and time-consuming.
- **Pilot Testing—**Continuous process testing should be conducted with small-scale filters or membrane units, using the

information collected from the previous sections. As noted, one or more treatment types may be tested simultaneously.

NOTES:

a. The protocol should include testing of a range of flow rates and chemical rates to ensure that the optimum operating conditions are found. However, the entire testing plan may not have to be completed if the bench-scale/jar testing provides an accurate starting point or if an acceptable solution is easily found. However, adequate testing should be completed to insure that the optimum conditions have been identified.

b. The duration of the testing will vary depending on the source water and the requirements of the regulatory agencies. For example, groundwater normally changes very little and a short testing period may be adequate. However, testing through several seasons may be required for surface water sources.

BENCH-SCALE/JAR TESTING

Bench-scale/jar testing should be performed to determine a starting point for the pilot testing program. Here again, the intent of this testing is to simulate the full-scale treatment processes. A brief outline for this procedure is given in the following sections. Appendix A has more complete information on the following steps.

1. Prepare stock solutions. Add various amounts of chemical to the jars. NOTE: Only one chemical at a time should be tested, until the optimum feed rate for it is established. Then, that amount can be added to all the jars, and the next chemical can be added in varying amounts.
2. Simulate a flash mixing step by stirring the jars rapidly for 30–60 seconds.
3. Determine the contact time necessary for the desired chemical reactions (oxidation or coagulation).
4. Simulate clarification (if required) by turning off the mixer and letting the jars settle for 2–3 minutes.
5. Simulate filtration by decanting the jars and running the supernatant through a vacuum filter. A vacuum filter is a necessary device for testing high quality raw water and for determining if the finished water quality goals can be achieved.

NOTES:

a. The optimum pH range for some coagulants such as alum is critical. For those cases, a buffering chemical may be necessary to arrive at the optimum point.

b. Records should be kept for all tests, which will provide an operational history for the plant. Historical records will be discussed in more detail in chapter 11 of this handbook.
c. Refer to Reference 1 for an extended discussion of bench-scale/jar testing. Refer also to appendix A for sample jar testing forms and procedures.
d. Verify these results for compliance with removal/treatment goals. Depending on the type of contamination being treated, the removal indicators may be different. Although not a complete list, several contaminants are discussed in the following section.
e. Typical contaminants:
   - **Iron and Manganese**—Lab test kits should be available on site for these contaminants. A spectrophotometer is recommended for more accurate results, not a simple colorimeter.
   - **Arsenic**—Accurate results for arsenic removal may require formal lab testing, which takes time. However, during bench-scale/jar testing or pilot testing, more rapid results are required. Iron frequently occurs with arsenic in groundwater or may be added to assist in the removal of arsenic. In either case, testing for iron removal may be used as an indicator for arsenic and is often acceptable for preliminary pilot test results.
   - **Turbidity**—Residual turbidity is easily measured by a spectrophotometer or bench-scale turbidimeter. However, the samples should have first been run through a vacuum filter as previously discussed.
   - **Color/Organics**—Dissolved color will normally pass through a filter, perhaps even a microfiltration unit. Properly coagulated color can then be removed by filtration. Removal efficiency can be verified by a spectrophotometer or turbidimeter, after chemical coagulation and vacuum filtration.

BENCH-SCALE/JAR TESTING SUMMARY

Sample jar testing forms and procedures are included in appendix A.

The bench-scale/jar testing procedure should not be performed until adequate lab instrumentation is available to evaluate the results, as previously noted. As a general rule, if jar testing is successful, the pilot filter study will be also.

There is no point in starting a pilot filter study until the bench-scale tests are successful and a testing plan (protocol) has been developed. Starting without either can significantly extend the testing period and waste time and money.

CHAPTER FOUR

Pilot Filter Composition

PILOT FILTER TYPES

The design of pilot filters often varies, depending on the contaminant to be removed and the desired finished water quality. Pilot filters can be of a variety of types, including the following:

Conventional

Figure 4-1 is an example of a completely assembled conventional pilot filter for granular media, including all the components described in the following sections, with all equipment mounted on a back panel, which has wheels for ease of movement.

Conventional media types may include multi-media (anthracite, silica sand, and garnet sand), dual media (anthracite and silica sand), or mono-media (coarse sand, gravel, or anthracite).

Figure 4-2 is a flow diagram illustrating the components used for the pilot filter column in Figure 4-1.

Conventional Pilot Filter Composition

Although some pilot filters operate by gravity, pressure filters are more common and compact, even if the full-scale plant will operate by gravity. For conventional filtration, a minimum 5-ft-tall clear column is recommended with a 6-in. inside diameter. Larger diameter pilot filters are recommended if available, because the wall surface (side wall) effect is less. Refer to Reference 1 for a discussion of *side wall effect*.

NOTE: The walls of a pilot filter are typically glass or plastic, and as such are "slicker" than full-scale filter walls and tend to have a greater tendency for short circuiting down the walls, thereby bypassing the media and resulting in poorer filtration efficiency. Therefore, the bigger diameter the pilot filter, the better the results may be. Full-scale granular media filters are often more efficient than pilot units for this reason.

Figure 4-1 Typical pilot filter

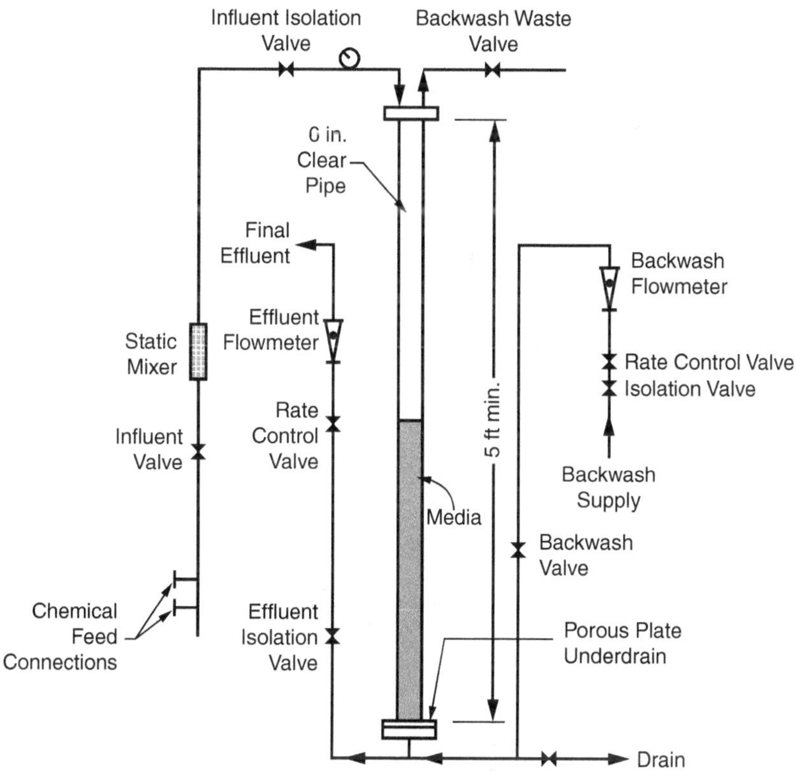

Figure 4-2 Conventional pilot filter flow diagram

Porous Plate Underdrain

As used in municipal type filters, a porous plate is a substitute for support gravel over some underdrain types. It consists of plastic beads heat welded together, with the intent of retaining filter sand without the use of the support gravel. Porous plates are used to cover the normal underdrain and to eliminate the height in the filter normally required for support gravel. The main underdrain should provide the water distribution capability, not the porous plate. A porous plate may actually be used as the underdrain in the pilot filter to eliminate the height required for support gravel, even if gravel is to be used in the full-scale plant. The small size of most pilot filters would also allow the porous plate to be used for distribution purposes.

A porous plate in the pilot filter also allows the use of an air scour cycle as part of the backwash system, which is preferred for most applications. Refer to Reference 1 for an extended discussion of air scour systems and use. However, the pilot filter underdrain and support may

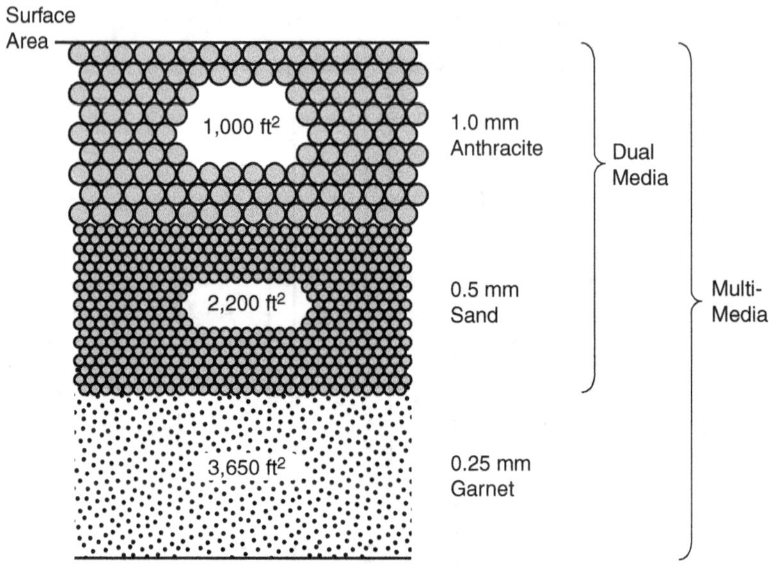

Typical mixed media load surface area equals 4,400 ft^2 (based on the use of 16½ in. of Anthracite, 9 in. of sand, and 4½ in. of garnet sand)

NOTES:

1. Dual-media and multi-media designs use a layered approach to filtration where coarse solids are removed in the upper layer and the finer solids are removed at the bottom.
2. Multi-media uses a small particle size and a high surface area to improve the capture of small particles.

Figure 4-3 Typical media types

be best, on larger units, if they represent the equipment and material to be actually be used in the full-scale plant, even if additional height is required. Some owners may, in fact, prefer it.

Filter Media

The filter media used must be compatible with the underdrain and/or support gravel for both the pilot study and the full-scale plant. For example, not all porous plate underdrain systems are compatible with all granular media types. A brief description of some conventional media types is contained in Figure 4-3.

Process Valves

The same valve arrangement should be used in the pilot as is intended for the full-sized plant. The valves can be manual or automatic. For temporary pilot filters, manual valves are most common, while automatic valves may be used for permanent units. The valves should include the same as will be used in the full-scale plant and typically include

- Influent on/off control
- Effluent on/off and rate control
- Backwash on/off supply
- Backwash rate control
- Backwash waste disposal
- Air scour supply

Flowmeters

Final effluent and backwash rate flowmeters should also be provided so that the actual rates used can be recorded. For temporary pilot units, pitot tube–type flowmeters may be used. For permanent units, flowmeters with an automatic transmitter may be desirable.

- **Pressure Gauges**—High quality pressure gauges and/or a differential pressure gauge for measurement of head loss are recommended.
- **Back Panel**—All equipment should be installed on a rigid back panel with the necessary piping and hose connections for influent, waste, effluent, and backwash purposes.

Greensand

The same type of pilot filter can usually be used for greensand as well as for conventional media. There are several types of greensand, but the requirements for pilot testing are similar. It should be noted that a minimum depth of greensand of 30 in. is recommended, with an anthracite cap, if suitable sized material can be found. Any anthracite should be sized so that it will sit on top of the greensand without excessive intermixing.

Microfiltration

Skid-mounted microfiltration pilot units (Figure 4-4) are available and may include

- A minimum of one membrane module
- All the necessary pumps, valves, and controls
- A compressor to operate pneumatic valves (if used)

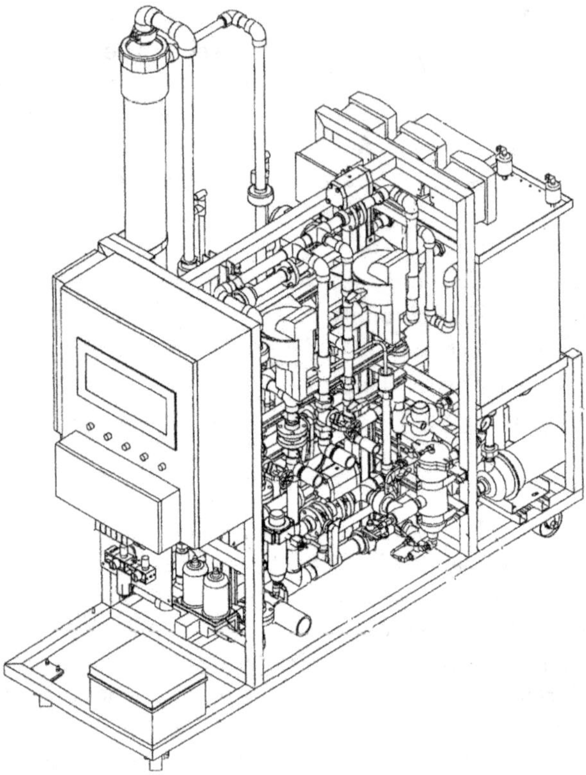

Courtesy of Pall Corporation

Figure 4-4 Skid-mounted microfiltration pilot unit

- Clean-in-place (CIP) equipment and chemicals to maintain and clean the membrane module
- All required on-line instrumentation
- Computerized controls

A flow diagram of microfiltration pilot equipment as supplied by the Pall Company is shown in Figure 4-5.

Note: The microfiltration pilot filter equipment previously described is provided with a complete operations manual, including installation instructions, and a troubleshooting guide.

Reverse Osmosis

The equipment for reverse osmosis is basically similar to that for microfiltration, except that higher pressures are used along with higher quality piping and valve materials.

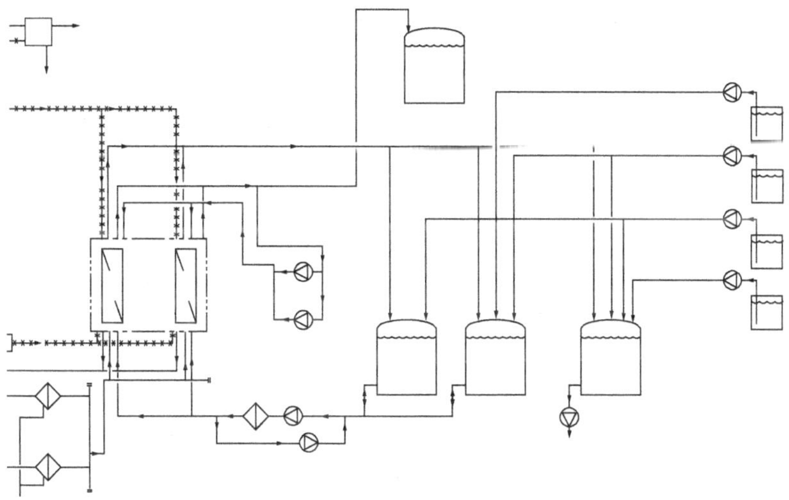

Courtesy of Pall Corporation

Figure 4-5 Flow diagram of the microfiltration pilot equipment

Slow Sand Filtration

Pilot filter units for slow sand filtration are not common and are typically built for the application, if they are required. As with all treatment types, the unit flow rate should be similar to that intended for the full-scale plant. Care should be taken to introduce the water into the filter in such a manner so as not to disturb the top of the sand.

PILOT FILTER EQUIPMENT SETUP

Many of the following recommendations may seem obvious, although there are installations that have not been set up properly.

- **Installation**—Make sure the equipment and chemical feed systems are plumbed correctly with materials compatible with the water to be tested.
- **Flocculation/Coagulation**—Flocculation/coagulation (if needed) will require detention time between the chemical injection point and the treatment unit. The amount of detention time will vary depending on the water temperature and the chemicals used. Some pilot units use a length of hose or piping for that purpose because it is easily modified.
- **Oxidation**—The oxidation of groundwater contaminants may only require minimal detention time and may be satisfied by the normal piping and fittings already used in the pilot unit.

- **Chemical Feed**—Chemical feed systems for temporary pilot units often use 5-gal buckets or plastic trash cans for mixing and feed tanks.
- **Dilution of Chemicals**—Dilution for the chemicals must be calculated for the size of the pumps, the pilot equipment operating rate, and the anticipated feed rate (mg/L). It is important that the chemical mix dilution be made as accurately as possible, as it has a major impact on the results of the testing, as well as the selection and sizing of equipment for the full-scale plant. Typical calculations of the chemical pump rate is as follows:

$$\text{chemical pump rate (gpd)} = \frac{\text{mgd (pilot rate)} \times \text{mg/L (chemical feed)}}{1{,}000{,}000 \times \text{(chemical mix dilution)}}$$

- **Piping Connections**—A fresh (treated) water supply should be located and provided for backwash purposes. A discharge point for treated effluent and backwash waste should be located and provided as well. An untreated raw water source with the proper flow for the pilot unit(s) should be located and provided.
- **Instrumentation**—As noted previously, the pilot filter testing should not be performed unless the appropriate lab instruments are available to provide fast results for all the required data. Once good results are available, samples should be sent to a certified laboratory for formal confirmation.

PILOT FILTER OPERATION

The pilot unit(s) should be turned on. The raw water flow and chemical feed pumps should be turned on as well.

NOTE: The initial setting for the chemical feed pumps should be as determined in the bench-scale tests.

Take samples for testing after at least one volume of water has passed through the pilot unit.

Process Adjustments (Flow Rate and Chemical Feed Rates)

It is recommended that chemical changes not be made until after processing several bed volumes of water, which are adequate to produce consistent results. If the unit is not producing the desired water quality, it is recommended that chemical changes be made at that time. However, only one chemical should be changed at a time for testing purposes, to accurately determine its effect.

NOTE: Generally, the water quality will not improve with longer runs without process changes, which may include chemical settings and/or changes in flow rate or flux rate.

Even if the initial water quality is good, testing should be continued until the optimum conditions are achieved and a confidence level is reached. If treated water quality is not good, the process flow rate or chemical feed rates should be adjusted (making only one change at a time) and testing should be tried again.

Record Keeping

Records of each run should be kept, either on paper or electronically, including chemical feed settings and unit flow rate. Keep a separate record of results any time any setting is changed.

Goals

The goal is to optimize both flow rate and chemical feed rates. High water quality, long filter runs, and the minimum chemical feed settings to achieve those conditions are the desired goals. These values will then be used in the design of the full-scale plant.

Formal Results

Once the on-site lab instruments indicate good results, representative samples of the treated water should be sent to a laboratory for formal confirmation. The laboratory results can then be included in the final report.

Consistency

When acceptable water quality is produced, the pilot equipment should be run as long as necessary for the operators to become familiar with the system and to determine if the filter run times are acceptable. As an example, if the water quality being produced is good but the filter runs are short, some process adjustments will be required followed by more testing to insure that optimum results have been reached. Also, each filter run should be similar in length to the others. If one is noticeably longer or shorter than the others, there is some problem that may have to be corrected and then reverified by more testing.

Example

The operation of a microfiltration pilot plant, which had never produced good quality water after two months of operation, was taken over by the author. The previous operators had attempted to operate the pilot plant on the same theoretical chemical feed rates the entire time without changing anything and without doing any bench-scale tests to determine the proper settings. The pilot testing failed because of the lack of a comprehensive testing protocol covering a range of flow conditions and chemical feed settings, and there were also errors in setting up the equipment.

The first step after taking over the pilot study operation was to conduct bench-scale/jar tests to establish the proper chemical feed rates, which took approximately two days once the tests were started. Meanwhile plumbing changes were made to the test unit. On the third day, the first day of operation, the pilot plant was producing high-quality water.

If the recommended complete testing procedure is followed, the pilot treatment should be successful in a very short time. Testing for verification can then be run as long as desired.

PILOT TESTING SUMMARY

The importance of bench-scale/jar testing is often overlooked due to the presence in many plants of streaming current monitors (SCM) and particle counters. However, bench-scale jar testing is highly important to identify the proper operating range of the plant quickly.

If problems occur while operating the full-scale plant, the pilot study report should be referred to in order to determine how those problems were resolved.

A permanent pilot filter unit is recommended for all larger treatment plants. These units have numerous uses including an early warning of treatability changes, operator training, and conducting chemical trials if testing new chemicals is desired. It is recommended that chemical trials not be conducted on the full-scale plant.

NOTE: Package treatment plants ordinarily do not have permanent pilot filters because of the short residence time in that type of equipment. The residence time is the length of time required for the water to pass completely through the plant. For a conventional plant, the residence time could be 2 hours or more, while a package plant may only require 20–30 minutes.

It is also recommended that the plant operators be involved in the pilot filter program to learn how the new full-scale equipment will operate. Training the plant operators in the use of a pilot filter is highly recommended so that they can conduct the tests and determine for themselves how the filter works. Hands-on training is always the best.

CHAPTER FIVE

Process Design Notes

The following process design notes are provided so that the operator may have a better understanding of how the plant should be laid out and organized, prior to the actual design process. It is not necessary that the operators be designers or that they have design training. It is necessary that the operators have a good understanding of the process issues and equipment with which they will be working.

BASIC DESIGN PHILOSOPHY

The design of the full-scale plant should be based on the pilot study and/or similar plants on the same source water, including unit flow rates, chemical processes, and so on. Also, the plant controls should be designed to implement the processes that were piloted and should be simple to operate. In addition, it is recommended that the pilot study report be included in the plant operations manual as a reference for the operators.

The designer should provide a plan designed to meet the clients' specific needs. If a complex operational system is needed, the plan should reflect this and not just a generic operation.

The clients should be involved during the entire design process for several reasons. They will know if the project meets their needs; they will have a better idea of the operational requirements; and they will be able to have input on issues relating to maintenance.

The clients need to "buy in" to the design procedure and treatment processes. *Buy in* means that clients have been involved in every phase of the design process and both understand and are in agreement with all the issues.

Operation

The clients should provide input on how they would like/need to operate the system. Again, a complete description of the proposed plant operation and the anticipated operational requirements should be provided to the clients during the design process in sufficient detail to allow them to properly understand the issues.

If the plant is remotely operated, it may need more operational alarms and automatic features. Some small plants may operate this way, but it is not recommended. Operators need to see what is happening. For example, there could be a chemical leak, which would need to be detected and repaired very quickly. If remote operation is necessary, a portable laptop computer is recommended for operators to carry when away from the plant, with full plant operational capability.

If the operators are on call or are part-time, the control issues will be similar to those discussed previously. Here again, part-time operators are not recommended when the public health is involved, unless the raw water quality is very stable and easy to treat. Funding for a full-time operator will be difficult for a small system. However, it should be noted that a small system has the same operating issues that a large plant has and the same health concerns.

When the operators are on-site full time, they can react to problems faster and more effectively. In this case, the controls can allow for more operator decision making, with less reliance on automatic controls, which is desirable in many cases, especially when the raw water quality is "flashy" and subject to rapid change.

Continuous operation is best, even if the plant has to be operated at a reduced flow rate. Frequent starts and stops are not normally good for treatability. Steady-state operation is best.

Frequent on/off operation can result in unstable flow in the plant, depending on the type of treatment equipment used. Variable rate operation can also result in inaccurate chemical feed changes, depending on the accuracy of the equipment. For example, flowmeters can be slightly out of calibration, and chemical feed equipment may not respond in a linear fashion, which can result in inaccurate chemical feed rates. In other words, a 10 percent increase in plant flow might only result in an 8 percent increase in chemicals, due to pluggage or fouling in the chemical pumps. For that reason, frequent cleaning and calibration of the chemical pumps are recommended. Therefore, an on/off type of operation is not recommended unless absolutely necessary. Even then, constant operator attendance is recommended.

A comprehensive discussion, or discussions, should be conducted with the plant owners, to insure that they understand completely what will be required for proper operation. Many issues are involved, and clients would do well to be prepared for the requirements for taking over and operating a new plant.

Maintenance Requirements

A periodic review of the plans during the design process can have a number of benefits for the owner/operators. For example, hose bibs can be located in multiple locations, as suggested by the operators, for cleaning and maintenance purposes. Adequate means for receiving and handling shipments of chemicals is another important issue. Adequate walkway space may need to be provided in equipment areas for clearance for a small forklift or pallet jack to move chemicals. Adequate space for access to equipment for maintenance purposes is another necessary item to be reviewed.

SUMMARY

The size and sophistication of the plant will obviously have a great impact on the design and the subsequent operation of the plant. It is, therefore, of great importance that someone with significant operational experience be in charge of the design, and that the clients are involved in the design process to increase their awareness of any operational or maintenance issues.

CHAPTER SIX

Operational Design Issues

There are obviously a number of design issues that can affect the operation and maintenance of the plant. Some are discussed in the following sections.

RAW WATER SUPPLY

The design of the raw water supply system can impact the operation of the plant. In addition, these operational issues are different for surface water and well water supplies.

Surface Water

Depending on the design of the raw water supply system, the raw water introduced into the treatment plant may vary from that of the raw water source itself. There have been instances where the raw water intake pulled silt off the bottom of the impoundment, making the plant influent dirtier than it should be. The same condition can occur if the raw water has been stirred up considerably close to the intake. If the plant operators are having difficulties with the raw water quality, it may be necessary to have the raw water intake evaluated to determine if it is picking up excess silt or other forms of contamination. In some cases, design revisions have had to be implemented to improve handling of the source water.

Example. One surface intake was constructed on a river, where excess silt had built up against the intake well, causing pumps to fail and large amounts of silt to be sent to the plant, which overloaded the solids handling system. A silt dam had to be built up against the intake well to keep the silt out but still allow water in.

Every aspect of a plant design must be carefully considered to minimize both operational and maintenance problems.

Well Water Supplies

Wells have their own peculiar set of requirements: they have to be sited so that there is no influence by surface water; they have to be operated at a rate that will not draw the water level down too low; and they have to be tested periodically for contamination that might be drawn in from other sources. Some other issues are discussed in the following sections.

Example. One plant treated both well water and surface water after they had been mixed together. A decision was made to separate the treatment for the surface water and well water. When that occurred, unusual raw water quality from the well source was observed, which had previously been masked by the surface water treatment. The problem was traced back to the well. Many wells, with water containing iron and manganese, require disinfection several times a year for iron bacteria. Apparently, disinfection had never been practiced at this well.

Another problem at this same well was discovered regarding the use of oil lubrication in the well pump. Approximately 1 gal of food-grade oil per week was used for that purpose. Over time, the oil had passed through the pump and had accumulated down in the well casing and coagulated, creating a large volume of cottage-cheese type solids. Because of the iron bacteria and the oil, a considerable amount of solids was generated in the well casing and sent to the treatment plant. The well casing was completely fouled, and the well had to be rehabilitated.

During the rehabilitation, it was also discovered that the well's screen was not designed properly for the fine sand at that location. Many soil types have a wide range of particle size, so that during development of the well, the coarser particles are drawn in toward the screen to provide a filter for the smaller particles. However, in this case, there were no coarser particles. The well was rehabilitated, but it was not possible to change the screen, and the well was scheduled for complete replacement.

Comments

These examples are given to demonstrate that operational problems can occur that are often beyond the operators' control. In these cases, the operators will have to seek design assistance. However, it is important for operators to be able to identify the source of potential problems, even though they may not be able to resolve these issues by themselves.

CHEMICAL FEED SYSTEMS

As a general rule, flexibility should be designed into the treatment plant, to the extent possible. Examples of flexibility in chemical feed systems are discussed in the following sections.

Duplex Chemical Feed Pumps

Duplex chemical feed systems are recommended. For example, a full size backup pump should be provided for every chemical system as a minimum. Then, the second pump will be available in case the first one fails or is off-line for maintenance. To keep them both in good operating condition and to even out the wear, the operation of the two pumps should be alternated frequently, daily or weekly for example.

Dilution Systems

Dilution systems should be provided for all viscous chemicals. Dilution water, added after the chemical pump, can result in more efficient delivery of the chemicals and reduced pumping pressure, in addition to better mixing at the point of injection.

Viscous/Hard-to-Mix Chemicals

Equipment for hard-to-mix chemicals is often highly automated and self-contained. Nonionic polymers are one such example. These polymers are provided as a powder and are highly viscous when mixed with water. The self-contained types of mixing equipment may produce very good results. However, open and accessible tanks are simpler and easier to maintain. For that reason, a mix tank/day tank system is preferred. Refer to Reference 2 for a description of chemical feed systems.

Multiple Chemical Feed Points

Multiple chemical feed points are recommended wherever possible. For example, it may be desirable to feed a polymer at the head of the plant as a coagulant aid. It may also be desired to feed the same or a different polymer in front of the filters. One treatment plant uses a cationic polymer as a coagulant aid and a separate nonionic polymer as a filter aid. Because of the peculiarities in the raw water at that site, that plant required the use of both polymers at the same time for effective treatment.

Chemical Response Time

The chemical pumps should be located as close as reasonably possible to the point of injection. The purpose is to reduce response time to changes in chemical rates. The longer the chemical delivery pipelines, the longer the response time.

The response time is the time it takes for changes in chemical feed dosages to actually reach the point of injection. If the chemical delivery pipeline is filled with 100 percent chemical (the same as in a chemical feed system with no dilution), changes in pump rate will be effective immediately. However, if dilution water is added to the pumped chemical, changes in the pump rate will affect the diluted chemical concentration, and the response time will be the time required for the new concentration to reach the point of injection.

NOTES:

- Whatever the response time, it must be included as part of the overall plant response time to any process changes. Process changes will not be complete until the new chemical feed rates have passed all the way through the plant. It is very important that additional chemical or flow rate changes not be made until the previous changes have had adequate time to pass entirely through the plant.
- Changes in flow rate only will affect the plant performance much more rapidly. However, in order to maintain the same parts per million chemical rates, the pumps will have to be changed in proportion to the flow rate changes.

Weather Protection

In the case of a large outdoor treatment plant, all chemical systems should be adequately protected from the elements, including direct sunlight temperature as well as normal weather conditions. A simple sun roof may be adequate in some cases, while complete enclosure and air conditioning may be required in others so as not to exceed the maximum temperature rating for the equipment or the chemical.

Example. One installation had a chemical feed system located outdoors in the direct sunlight. Summer conditions were such that the direct sunlight temperature was close to, or exceeded, the design capability of both the pump and the piping systems. Numerous failures occurred including rupture of the piping and failure of the chemical pump diaphragm.

In another case, a polymer system was inside a room that had a roof but the sidewalls were open to the atmosphere and there was no air conditioning. Ambient air temperature exceeded that of the rated storage temperature for the chemical. As a result, the chemical was ruined, and it coagulated into a solid form that fouled the delivery pipeline.

Methods of Injection

It is recommended that all chemical injection into a pipeline be done by way of an injection quill as shown in Figure 6-1. Chemicals injected in this manner will be much better distributed into the flow stream. In addition, the injection quill helps protect the pipeline material, which

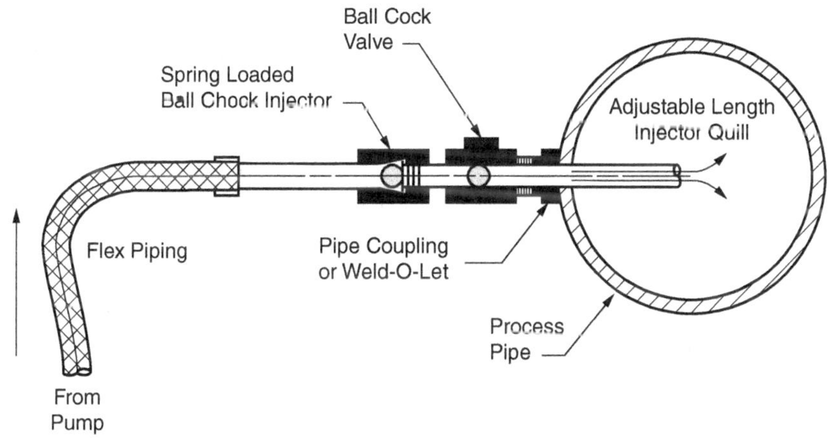

Figure 6-1 Injection assembly

is essential with metallic pipe. It may be obvious, but any wetted pipe or tubing materials should be compatible with the chemicals used. Other types of injection and chemical mixing are discussed in the following sections.

Examples. One site fed gaseous chlorine into the process piping through a brass fitting, which rapidly corroded and failed. Another plant injected a corrosive chemical into an ordinary grade of stainless steel piping and fittings, which also failed. In both cases, the use of an injection quill with plastic fittings would have resolved or greatly mitigated the problem.

Process Water Mixing

It is essential that chemicals be adequately mixed (rapid or flash mixing) into the process water stream as quickly as possible, especially when using coagulants or polymers. A total energy input of 3 ft of head loss is recommended for this purpose. For more specific recommendations, the chemical equipment manufacturers should be consulted for their recommendations.

NOTE: An important factor in the design of rapid or flash mixing is the anticipated variation in plant flow to be expected throughout the various seasons of operation. Mixing is accomplished in one of several ways.

Hydraulic jump or weir. Either a hydraulic jump or flow over a weir can be used for rapid or flash mixing. However, when using either, the effect on mixing of changes in flow rate should be carefully considered.

If either is used, the designer should anticipate any potential problems that may be caused by entrained air resulting from the surface agitation.

Static mixers. Static mixers typically consist of a series of vanes inside the process pipeline. The intent of the vanes is to promote intermixing of the process water. However, the designer and operator should both be aware that the mixing intensity provided by static mixers varies with the process flow rate.

For example, if the necessary head loss is provided at the maximum design flow, the mixing intensity may be reduced by 75 percent if the process flow rate is reduced by 50 percent. Therefore, it is extremely important to determine in advance the anticipated flow rate conditions under which the plant is expected to operate, and to make sure that adequate mixing will be provided under all operating conditions.

In another example, package treatment plants may operate in an on/off mode at or close to the design rate most of the time. In that case, a static mixer might work well. However, larger plants may be operated at different flow rates, requiring different mixing. In either a package plant or a full-scale plant, the design must consider the anticipated plant operational modes.

Mechanical mixing. A mechanical mixer consists of a motor outside the pipeline that drives a shaft through water-tight bearings to a rotating propeller in the process stream. The amount of agitation is constant and is usually designed for the maximum flow. At reduced plant flow conditions, the actual mixing intensity is greater than designed.

Recommendation. Unless the plant is intended to operate at the same flow rate in an on/off condition, a mechanical mixer is recommended for a more flexible design, although a hydraulic jump may work as well if the plant design allows for the use of one.

Troubleshooting. If adequate rapid or flash mixing is not provided, the treatment efficiency between multiple filters may be different. At one treatment plant, one filter received more chemical than another because of the lack of adequate mixing at the point of injection. If treatment irregularities occur between filters or process trains, the adequacy of mixing after chemical feed addition should be investigated.

CHEMICAL DETENTION TIME

In many plants, the residence time (or detention time) of the chemically mixed water in the pipeline between the point of injection and the first stage of treatment can be critical. The necessary time can vary widely, depending on the source water, the type of treatment, and the water temperature.

Surface Water

With surface water, it is often necessary to coagulate the raw water with chemicals. The time required for that process to take place is dependent on the type of chemical used and the water temperature. The construction of the plant must allow for the necessary detention time, prior to the next treatment step. These issues should have already been evaluated and determined during the pilot study process. If there has been no pilot study, a very conservative design may be required.

Well Water

Typical contaminants in well water are iron, manganese, and arsenic. When removing these contaminants, an oxidation step is usually required. Chlorine and/or potassium permanganate are often used for this purpose. In that case, very little detention time is usually required for the oxidation process to occur. The normal piping system between the point of chemical injection and the filters is typically adequate for that purpose. However, this also should have been determined during the pilot study program.

Rapid or flash mixing of some sort may be necessary to thoroughly mix the chemicals in the water, to ensure the efficiency of the oxidation process if the detention time is short. Numerous pipe fittings between the point of injection and the filters may also be adequate for mixing purposes. Again, these issues must be carefully evaluated during the pilot study and the design process.

NOTE: Many operational design issues can be rectified after the initial construction is complete. However, the lack of adequate contact or detention time is difficult to change at that point. Therefore, this factor must receive considerable attention during the design process.

CHAPTER SEVEN

Process Design

Process design includes a number of factors such as selection of the basic treatment process including the plant's water chemistry (should be done as part of the pilot study and preliminary design); selection of the operating components such as pumps and valves; selection of the underdrain, filter media, and/or membrane systems; and the selection of and location of the instrumentation. The last step is to develop the control system that operates the plant.

TREATMENT PROCESSES

The following discussions focus on designing the basic processes of a hypothetical water treatment plant. The intent is to determine the instrumentation, valves, and pumps to be used, and where the individual components are to be located. During this process, a flow diagram will be constructed, which is the first task in the design process. Preparing a flow diagram is also very important in establishing the operational procedures to be used in the plant. Understanding the flow diagram is also highly important for the operators.

For this discussion, it does not matter which type of plant is used.

- Direct filtration (chemical feed followed by filtration only)
- Package plant (a complete premanufactured treatment plant)
- Conventional (complete treatment including the following processes)
    - Chemical feed (coagulation, filter aid, etc.)
    - Sedimentation (clarification to remove heavy solids)
    - Filtration (removal of fine solid particles with granular media)
    - Disinfection
- Reverse osmosis (membrane filtration, usually for seawater, may need pretreatment)
- Ultrafiltration (membrane filtration)
- Microfiltration (membrane filtration)
- Special medias (greensand, etc.)
- Or any combinations of the above

ASSUMED TREATMENT PROCESSES

This discussion will focus on a hypothetical conventional plant, including the components listed above for that type. The operational philosophy is more important than the plant type. Once that is understood, it is relatively easy to modify a flow diagram to fit other treatment types. The complexity of process design is in determining how to control valves, pumping systems, and multiple treatment trains while taking some of them off-line for backwash and then turning them back on, all the while maintaining the design flow rate. Valve speed even becomes important in maintaining proper control.

COMMON CHEMICALS

Selecting the chemicals needed should also be done in the pilot study/preliminary stage. The necessary pumps, tanks, piping, mixing equipment, and other related components are then selected to be compatible with the chemicals and are sized to meet the needs of the plant. Refer to Reference 2 for an extended discussion of these issues.

The following are some of the typical chemicals used in treatment plants. First listed are coagulants and filter aids:

- Alum—$Al_2(SO_4)^3 \cdot 14(H_2O)$
  - Polyaluminum chloride—PAC
  - Ferric chloride—$FeCl_3$
- Polymers (filter aid or primary coagulant)
  - Cationic (liquid)
- Often used as a filter aid.
  - Nonionic (powder)
- May be used as a filter aid or as a primary coagulant.
  - Anionic (less common)

Common oxidizers (used as disinfectants and/or oxidation of dissolved metals) include the following:

- Potassium permanganate—$KMnO_4$
- Chlorine
  - Gaseous—Cl
  - Liquid—NaOCl
- Ozone—O_3

Common taste-and-odor control chemicals include:

- Powdered carbon (C)
- Potassium permanganate—$KMnO_4$
- Ozone—O_3

Common buffering agents (used to raise pH and to add alkalinity) include:

- Lime
- Soda ash
- Sulfuric acid—H_2SO_4

Other common disinfectants include:

- Gaseous chlorine—Cl
- Liquid chlorine—NaOCl
- Ozone—O_3
- Ultraviolet light—UV
- Chlorine dioxide—ClO_2

NOTE: Each of the chemicals listed has specific design and operational needs. For example, the use of alum is pH dependent. In addition, many of these chemicals are corrosive and require special handling and piping materials.

PROCESS FLOW DIAGRAM

The water chemistry for a plant is tailored for each specific site, treatment type, and the contaminants that are present in the raw water. The instrumentation and controls are then designed to implement the water chemistry, all of which should be illustrated on a flow diagram. The completed diagram then becomes the basis of design and may include some of the equipment listed in the following sections.

Basic Processes

The basic treatment processes include raw water intake pumps, chemical feed pumps and mixing equipment, instrumentation, clarifiers, filters, process flow control, and so on. These items are discussed in more detail in the following sections.

Process Flow Control

Plant flow control starts with the raw water intake whether pumps are used or not. On/off or proportional control may be used, depending on how the plant is operated. Larger plants may use on/off control of multiple pumps and perhaps one variable speed pump for more accurate control.

Flow splitting is required when there are separate treatment trains. Separate flowmeters and flow control valves are usually required for each treatment train for flow splitting purposes. Weirs may also be used for flow splitting purposes.

Accurate backwash flow control is necessary to insure adequate, proper cleaning of granular media (if used). The same is true for membrane systems. Flowmeters, flow control valves, and/or variable speed pumps may be used for this purpose

Accurate clearwell effluent flow must be monitored and recorded as a measure of product water sent to the distribution system. The clearwell effluent flow may be either by gravity or pumps. In either case, on/off or proportional flow control may be used. Modifying or controlling the effluent flow in any way requires that the raw water flow be adjusted accordingly.

Nearly every water system uses some type of chemical feed equipment. Water systems also need and use mixing after injection and either on/off or proportional flow control. Smaller systems tend to operate more in an on/off mode using manual speed controls, without any proportional or automatic variable speed control. Most other basic issues concerning chemical systems will be similar whether the system is large or small.

Raw Water Instrumentation

Raw water instrumentation is used for determining if adequate raw water is available for running the plant, for controlling the flow of water into the plant, and also for monitoring the water quality. Different types of commonly used instrumentation are discussed below.

- Well or surface impoundment level controls are used to insure that adequate raw water to run the plant is available. The level probes used for this function perform essentially the same service but may be of different types; one type having to go down a well casing, and the other usually having a wider surface to measure.
- Monitors for pH/temperature are typically for surface waters only because well water tends to stay at a relatively uniform temperature. The data from the instrument are used to monitor any changes in raw water quality and to assist in determining any process changes that need to be made.
- Turbidity monitors are also typically used for surface water only. The turbidity of well water does not usually change unless there is some sort of drastic failure. Turbidity monitors are also used to measure raw water quality and to assist in determining any process changes that need to be made. In worst-case conditions, the turbidity monitor may shut down the plant if the raw water becomes too dirty.

- Particle counters are also typically used only for surface water. A comparison with the final effluent particle counts helps determine the efficiency of the treatment process.
- Silt density index (SDI) monitors are usually used for reverse osmosis applications to determine the effectiveness of the pretreatment processes.
- Conductivity is also used to determine the effectiveness of reverse osmosis.
- Alkalinity and/or hardness monitors are used for both surface water and saltwater applications to assist the operators in determining if any process changes are needed, especially in regard to coagulation.
- Flowmeters are used for all applications both to monitor the flow and to set the flow at the proper rate.
- Iron and manganese monitors (manual or automatic on line) are used primarily for well water to insure that water quality goals are being met.
- Silica monitoring is not common but may be used to verify the water quality for boiler feed (not common) applications.

NOTE: Automatic in-line monitors are recommended for all applications (if available).

Chemically Treated Water Instrumentation

Instruments for chemically treated water are located after chemicals have been added to the raw water and after mixing.

- A pH monitor is used in this location to verify that the coagulant dosage used is correct. It is especially necessary if a pH-sensitive coagulant, such as alum, is being used.
- Streaming current monitors (SCM) are also commonly used in this location. They monitor the particle charge in the water to insure that the water has been properly neutralized for coagulation.
- A chlorine residual analyzer may or may not be provided at this location, depending on the type of treatment being used. For example, if organics are present in the raw water, chlorine should not be added until after their removal in the clarification or sedimentation process.

NOTES: An SCM is not normally provided for well water supplies. A pH meter may not be required for all applications, but is recommended.

Optional Clarification or Second-Stage Monitoring Instrumentation (Primarily for Surface Water)

- A turbidity monitor is often provided (and may be required) after clarification and before filtration to verify the efficiency of solids removal to that point.
- Particle counters may be provided at this location for the same purpose as a turbidity monitor, as previously discussed.

Filter Effluent Instrumentation

Instrumentation is necessary and required after treatment to verify that the water quality meets the regulatory standards before being sent to the distribution system.

- A turbidity monitor is the primary device used to prove that the water quality meets or exceeds the standards for treatment.
- Particle counters are also used to measure treatment efficiency and can be used, in some cases, to determine if one of the treatment trains is not as efficient as it should be.
- An SDI monitor (manual or automatic) is commonly used to verify that the pretreatment efficiency is adequate before reverse osmosis treatment. It is not commonly used in conventional treatment.
- Flowmeters are used to set the proper operating rate for each treatment train, as well as for the complete plant. They are also necessary for measuring the amount of water that has been sent to the distribution system. There are many types of meters available, and the best and most accurate should be used. Refer to Reference 1 for a more extended discussion of flowmeters.
- A head loss monitor/transmitter is recommended for each granular media filter (if used) to help determine when the filter needs to be backwashed. A head loss monitor can also be used to help troubleshoot filters and determine if they are operating properly.

Clearwell level probes may be used in the clearwell for several purposes including the following: monitoring and controlling the level to insure adequate volume for disinfection contact time (CT), insuring an adequate volume is available for backwash purposes, and controlling the final effluent pumps (if used).

A final pH adjustment monitor (if required) is used to verify that the final product water is produced at the proper pH level to reduce or prevent corrosion.

Final Effluent Instrumentation

The final effluent instrumentation is located after the clearwell and after the final effluent pumps (if used) to verify that the product water was not contaminated after treatment or while in the clearwell.

- Turbidity monitors, pH monitors, SDI monitors, and particle counters have been discussed previously; however, they are also part of the final effluent instrumentation.
- A chlorine residual analyzer is required in this location to insure that the proper dosage of chlorine is added. Too much or too little will sound an alarm indicating to the operators that some change is necessary.
- A conductivity/hardness monitor is used typically for reverse osmosis treatment to verify that the pretreatment is run efficiently.
- Fluoride is often added to potable water. If so, it may be added in the clearwell after treatment. Installing an analyzer in the final effluent piping will allow the operators to verify that the proper dosage is used.

NOTES:

a. The above list of typical instrumentation is indicative of that which might be needed in a treatment plant and is not intended to be a complete list. In the preparation of a flow diagram, the instrumentation actually needed by the selected treatment process can be taken from this list, with others added as needed.

b. All data from the above instrumentation should be displayed on a computer screen for the operators' benefit.

c. Again, it is recommended that automatic in-line instrumentation be provided for all applications (if available).

SUMMARY

The intent of this chapter is to acquaint the operators with the various plant process components and how they are selected and located. The next chapter will locate those items in the flow diagram, which will then be the basis of design for the plant processes and equipment.

CHAPTER EIGHT

Preparation of a Process Flow Diagram

In this chapter, a process flow diagram will be developed for a hypothetical treatment plant using the instrumentation and other equipment described in chapter 7. Even though the treatment processes may vary from one site to another, preparing a flow diagram with the associated instrumentation and equipment is similar for most plants and should be the first step in the design process. The intent is for the design process to implement the flow diagram. This information is provided for operators to assist them with understanding the processes and the function of the various types of instrumentation and monitoring devices. It is recommended that all operators be familiar with the flow diagram for their plant.

The equipment and instrumentation are added to develop the flow diagram in a series of stages, in order to illustrate the design process. It should be noted that the selection of instrumentation and equipment in the stages shown do not necessarily have to be done in that order. The order of the stages is selected in order of treatment for convenience only.

STAGE 1—THE BASIC PROCESS

For our purposes, the basic treatment process is assumed to be as shown in Figure 8-1. Although a conventional treatment system with flocculation and sedimentation is shown, the flow diagram is similar for pretreatment followed by microfiltration or reverse osmosis. In this figure, filtration is followed by a clearwell with final distribution pump(s) and also includes a backwash system.

Stage 1 of constructing the flow diagram includes the instrumentation and equipment used for controlling the raw water supply and the finished water discharge. Each of the basic components added in this step is discussed in the following sections.

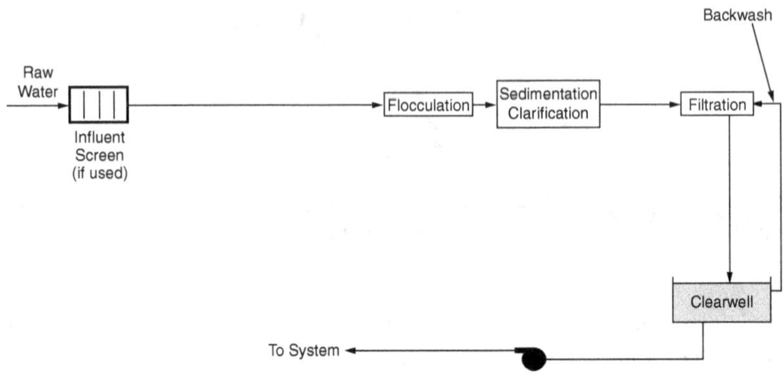

Figure 8-1 Process flow diagram, Stage 1: Basic process

Raw Water Supply

The raw water supply may be pumped from one or more wells, gravity flow from an impoundment, or one or more pumps from a surface source (either freshwater or seawater). In all cases, flow control and water quality monitoring are needed. When multiple pumps are used, whether from wells or a surface source, flow control becomes more complex as discussed in the following sections.

Flow and Pressure. When using multiple raw water pumps from either type of source, the pumps should have adequate pressure at full capacity as well as at reduced flow.

In the case of variable frequency (speed) drive (VFD) motors, the pumps have to produce the design pressure at the minimum flow, which means that excess pressure may be produced at high flow. Some process may be required to dissipate the excess pressure. Refer to Reference 2 for an extended discussion of VFD motors.

On/off control. If a pumping system is operating, care should be taken in turning the pumps on and off to mitigate pressure surges. Pressure surges can cause a number of different operational problems in the plant, including short-term water quality issues.

Finished Water Discharge

The finished water discharge includes clearwell design and capacity, level control, disinfection contact time (CT), and any required chemical feed and mixing.

Clearwell. The clearwell is intended, in many cases, to provide chlorine contact as well as to act as a pump well for the distribution pumps. There are several issues regarding the design of a clearwell including level control, chemical mixing (if post-filtration buffering is required), and the approach velocity to the pumps.

Level control and capacity. There can be considerable variations in the clearwell level, depending on changes in the plant operating flow rate, the operation of the filters, water required for backwash purposes, and the time required for the distribution pumps to turn off or come on line. For example, VFD pump motors (if used) may take 1 or 2 minutes to come up to full speed. The design of the clearwell is also an important factor. In some cases, the clearwell level is maintained at a constant level by a weir. Water then flows over the weir to the pump well where the water level could vary considerably depending on the operation of the pumps.

Therefore, it is common for pump wells to have a minimum 5-minute to 10-minute detention time (capacity), to account for these variations. In addition, the level and pump speed controls need to react very quickly. Refer also to Reference 2 and the pump manufacturers' recommendations.

Disinfection contact time. When the clearwell is used for chlorine disinfection, the contact time (CT) requirements will determine the minimum volume. Weirs may be used to maintain a minimum volume in the CT chamber, with the excess flowing over the weirs and into a pump well as previously noted. In that case, an additional 5 to 10 minutes of capacity may have to be allocated in the clearwell, for a separate pump well.

Chemical mixing. Chemicals may be added in the clearwell for disinfection (chlorine, etc.) and/or buffering purposes (alkalinity pH adjustment). When either is done, care must be taken to provide adequate mixing to disperse the chemical uniformly through the entire volume of the clearwell.

Example. At one treatment plant, lime was added to one corner of the clearwell. Because of the lack of adequate mixing, the pH measured in the clearwell effluent fluctuated widely, making the lime feed rate difficult to control.

Because of its high concentration, lime is fed as a slurry, with the intent of mixing and dissolving it in the flow stream before it can settle out. In this case, a large amount of lime settled out in the clearwell and created a mound over 6 ft high. The chemical settled out before it could dissolve, due to lack of adequate mixing.

NOTES: Chlorine disinfection is more effective at low pH. Therefore, maintaining a lower pH may result in a smaller clearwell. Refer to the standard USEPA CT tables for the required detention time.

Pump Approach

In any type of raw or finished water pump well, it is necessary for the process flow to approach the pumps in a smooth and uniform manner and without turbulence. Severe operational problems can occur in the pumps if there are problems with the approaching water.

Example. At one installation, the pump capacity of an older wet well was increased 100 percent without consideration to the pump approach velocities, and it resulted in significant pump operational and maintenance problems. Doubling the process flow had the effect of doubling the velocity of the water and creating turbulence as it approached the pumps, causing cavitation and excessive wear. The pumps were having to be replaced or rehabilitated every 2 or 3 years as a result.

NOTES: For assistance in pump well design, refer to the pump manufacturer, pump design manuals, and/or information contained in Reference 2. These comments also apply to the design of raw water intakes.

STAGE 2—FLOW CONTROL

Flow controls are added to the flow diagram in three locations in this stage, as shown in Figure 8-2: on the raw water, on backwash supply, and on finished water. In each case, the equipment consists of a rate control valve and a flowmeter. The controls' software (programming) modulates the rate control valve in order to achieve the desired reading on the flowmeter, as set by the operator. Programming software is a separate issue and is not discussed in this handbook.

Raw Water Flow

The raw water flow rate is often used to flow pace the chemical feed systems (recommended). In other words, if the operator changes the plant raw water flow rate, the plant controls automatically change the chemical pump rates proportionally so that the same chemical rate is used at the new plant operating rate.

Backwash Flow

An accurate flowmeter is required to monitor the backwash flow. It is important for granular filter media and even membrane systems to be washed at the appropriate rate. Too high of a rate can result in the loss of media, while too low a rate can cause fouling of the media due to inadequate cleaning. It should be noted that the desired backwash flow rate is temperature dependent and should be adjusted accordingly. Refer also to Reference 1 for a more extended discussion of backwash rates.

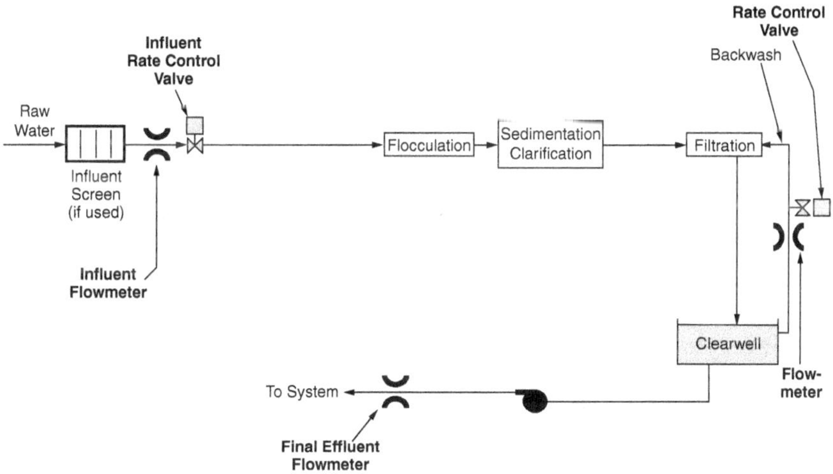

Figure 8-2 Process flow diagram, Stage 2: Flow control

Final Effluent Flow

The final effluent flow rate is the actual amount of treated water delivered to the distribution system. The final effluent flow rate may also be used to flow pace the final effluent chemical feed systems (if required).

Flow Control Valves

For flow control purposes with a low differential across the valve, electrically actuated butterfly valves are preferred that meet AWWA specifications.

Actuators. Electrical actuators are recommended for accuracy, for repeatability, for the available diagnostics, and for compatibility with computer controls. The gear drive on an electric actuator is much smoother than pneumatic or hydraulic units, which tend to be somewhat irregular and jerky in their movements. The manufacturer should be contacted if additional information is required on these issues.

Maximum differential pressure. Flow control valves modulate the flow by opening or closing as required to obtain the proper rate. The partially open valve then creates an amount of head loss as water passes through. The maximum differential head loss should be limited across a flow control valve of this type. Pressure gauges on both sides of the valve are recommended for monitoring the pressure loss.

The manufacturer should be contacted for differential pressure recommendations. As a general rule of thumb, the maximum differential pressure across a butterfly valve should be limited to 30 psi or less, preferably much less. A high differential pressure can cause cavitation with the associated vibration and valve wear. If greater differential pressures are required, a specialty valve of a different type may be required.

General flowmeter comments. Many different types of flowmeters are available from several different manufacturers, and it is necessary for operators and designers to know the operating characteristics of each. Operators and designers should visit other plants to see and hear what other operators like and dislike.

Entrained air and turbulence. Although some manufacturers claim that their flowmeters are not affected by entrained air and turbulence, entrained air should be eliminated to the extent possible, and flowmeters should not be installed close to pipe fittings, valves, or other piping components that can cause turbulence. Refer also to Reference 1 for an extended discussion of air removal.

Full piping. The pipeline and flowmeter should be full of water when either out of service or operating. Again, some flowmeters may be able to operate partially full, but it is better to not take chances if at all possible. Turbulence in partially full pipes may also be a problem.

Accuracy. An accurate flowmeter is desired to achieve proper operation of the plant. For example, if the flow measurements are not accurate, it may be difficult to obtain the proper chemical feed rates.

The flow inside a pipeline is seldom uniform across its width, especially when close to an elbow or other fitting. Normally, the highest velocity is expected to be in the center of the pipeline and lower velocity next to the side wall. However, the point of highest flow may actually be off center, with irregular velocities across the width of the pipeline. As a result, obtaining the best and most accurate flowmeter is recommended.

Recommendations. A summary of key flowmeter recommendations includes at least:

- Meters with no center bodies or obstructions in the flow path are recommended. Meters with these features may be difficult to keep clean and in proper operating condition.
- Magnetic flowmeters with multiple sensor/transmitters around the circumference of the meter body are preferred. Some of this type are able to operate with partially full pipelines and may also be able to tolerate some turbulence. Even so, eliminating entrained air and turbulence close to the meter is recommended for a conservative design and in order to help obtain the best data possible.

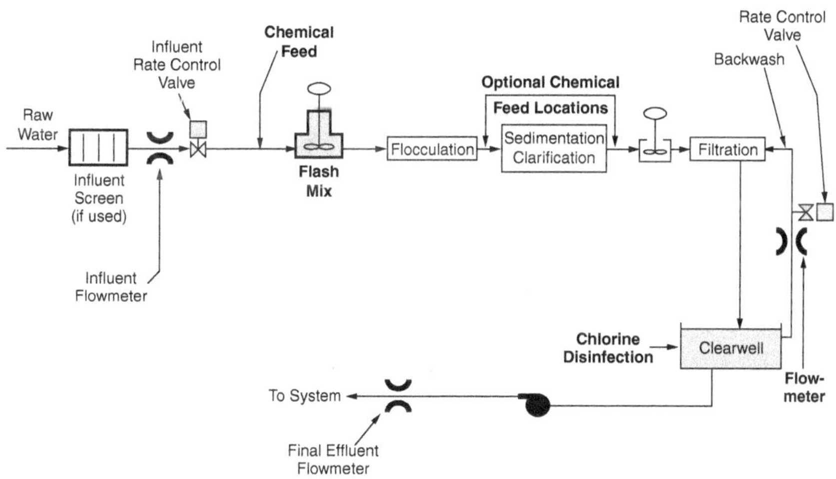

Figure 8-3 Process flow diagram, Stage 3: Chemical feed and mixing

STAGE 3—CHEMICAL FEED AND RAPID MIXING

Chemical feed systems are included on the process flow diagram in this stage. The chemical feed injection points, shown in Figure 8-3, are located after the raw water rate control valve, before and after sedimentation/clarification, and in the clearwell for this assumed treatment plant. This figure is for discussion purposes only. The actual chemical feed injection points will vary according to the needs of the process. More chemicals and more feed points may actually be required, as determined by pilot testing.

As stated previously, the conventional treatment processes shown might consist of pretreatment followed by membrane filtration. In either case, the chemical feed system requirements may be similar.

Chemicals Used

The actual chemicals used will vary according to the needs of the raw water and treatment processes, as discussed previously. They may be any of those mentioned in the previous section or combinations thereof, as determined by pilot testing.

It should be noted that duplex pumps may be provided for each application and at each feed point. They may also have proportional feed rate control as needed for the operational philosophy of the particular plant.

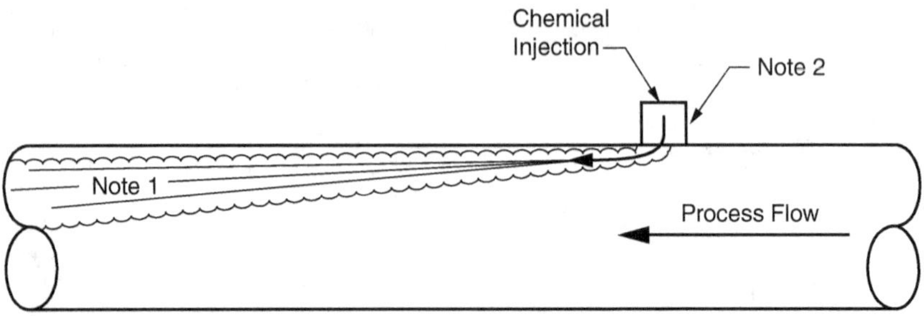

NOTES

1. The chemical tends to stay close to the pipe of conduit side wall. A considerable length of the pipe or a number of bends or pipe fittings are required to achieve complete mixing.
2. Injecting the chemical directly into the side wall of the pipe of conduit can lead to rapid corrosion, depending on the chemical used.

Figure 8-4 Direct chemical injection

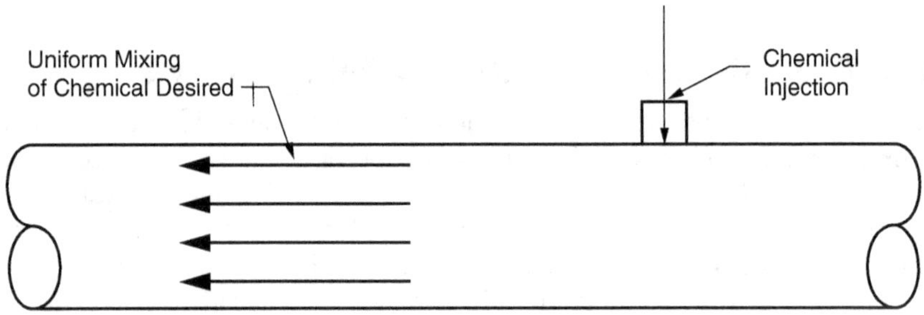

NOTE:
The intent of injecting a chemical is that it be thoroughly and uniformly mixed with the process flow as quickly as possible.

Figure 8-5 Desired mixing

Chemical Mixing

It should be noted that chemical mixing, of some type, is normally required after each injection point. The purpose is to evenly distribute the chemicals in the process stream. Inadequate mixing can result in the inefficient use of chemicals and uneven distribution of chemicals to the various stages of treatment. Figures 8-4 and 8-5 illustrate what the chemical addition may be like, with and without mixing. Mechanical in-line mixers are shown in Figure 8-6 for this purpose.

In-line mixer. An in-line mechanical mixer is shown where chemicals are added to the raw water. It has an electric motorized propeller in the pipeline that provides a constant amount of mixing at all times.

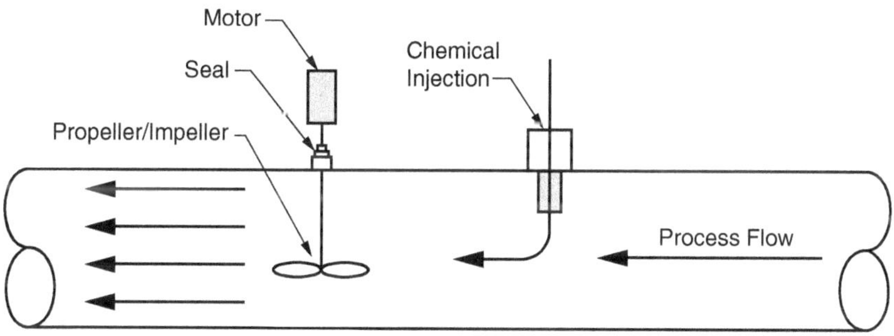

NOTES
1. Recommended energy input should be approximately 3 ft of headloss.
2. Motor size and mixing intensity vary depending on pipe size, process flow, and the chemicals being used.
3. Provides the same power input at any flow.

Figure 8-6 Mechanical flash mixer

Such mixers usually have ports in the mixer body for chemical injection. Figure 8-6 is a diagram of a mechanical mixer.

Open/gravity mixing. At one treatment plant, the flow from two treatment trains was combined in one small compartment prior to splitting to six filters. A large slow-speed chemical type mixer was used in that compartment, after the clarifiers, for mixing a filter aid polymer in the process water prior to filtration. The intent was to provide uniform dispersion of the chemical to all the filters.

Clearwell mixing. No mechanical mixing is shown in the clearwell in Figure 8-3. Chlorine and any other chemicals required at this location are typically added to the inlet piping to the clearwell, although mixing or adequate agitation is still required, unless the turbulence from the water flowing into the clearwell provides this function. Mixing at this and all other locations should receive considerable attention during the design process. Baffling is often added to the clearwell for CT purposes.

Chemical injection. It is recommended that all chemical injection into piping be either made through the body of the mechanical in-line mixer (if used) or by using an injection quill (Figure 6-1). Static mixers also have ports for chemical injection purposes.

STAGE 4—RAW WATER INSTRUMENTATION

Instrumentation is added at Stage 4 for monitoring of the raw water conditions. The intent is to monitor the quality of the raw water in

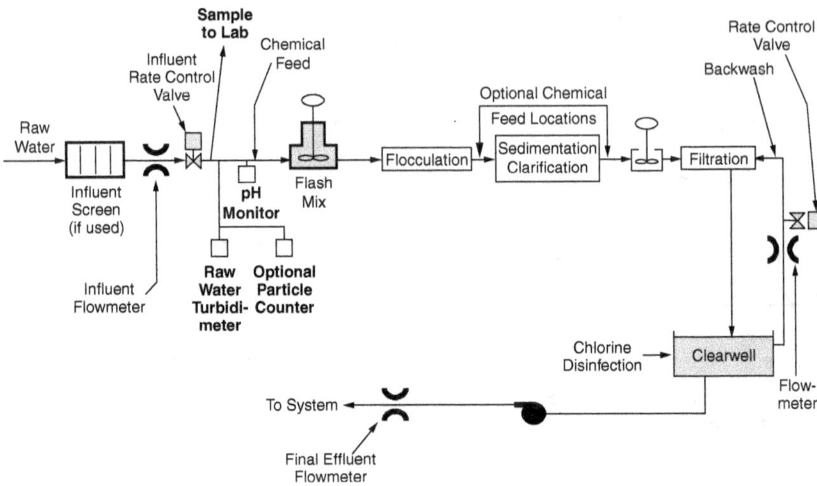

Figure 8-7 Process flow diagram, Stage 4: Raw water instrumentation

order to treat it properly. Also, changes in the chemical feed rates may be required if there are changes in the raw water quality.

Instrumentation

The instrumentation may include an SDI monitor as well as those shown in Figure 8-7. All this instrumentation should be installed prior to any chemical feed. It should be noted that the actual location of the instrumentation should be well in front of the chemical injection points in order to not be affected by them.

Laboratory Sample Water

For the operator's convenience, it is recommended that a sample of the raw water should be piped to the plant laboratory. A continuous flow of fresh raw water is a great benefit to the operators for use in bench-scale/jar testing, as well as for calibrating instruments. The pH of the raw water can change with temperature if a sample of water is allowed to sit for any length of time before testing. Therefore, a fresh and continuous source of raw water is desirable.

NOTE: Other sample water supplies will also be recommended in this handbook for use in the plant laboratory.

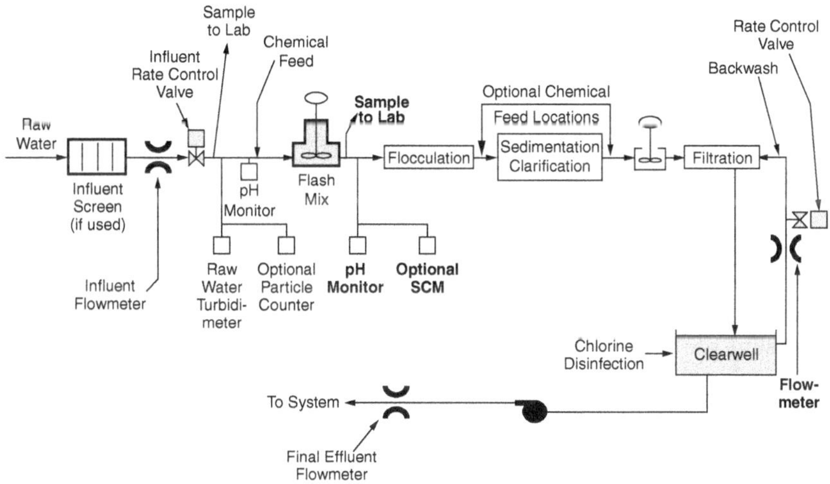

Figure 8-8 Process flow diagram, Stage 5: Chemically treated water instrumentation

STAGE 5—CHEMICALLY TREATED WATER INSTRUMENTATION

The next set of instrumentation is included after chemicals are added to the raw water and after flash mixing. The instrumentation may consist of a pH monitor and a streaming current monitor (SCM), as well as others (Figure 8-8).

pH Monitor

The purpose of a pH monitor at this location is to verify that the target pH for coagulation has been achieved, assuming that a pH-sensitive coagulant is being used. Other coagulants, such as polyaluminum chloride derivatives, may not be as pH sensitive. However, having the data is still desirable.

The target pH for coagulation (if used) should be determined by bench-scale/jar testing and/or an SCM. Even if an SCM is used, periodic verification of the SCM set point by bench-scale/jar testing is recommended.

Streaming Current Monitor

The use of an SCM is highly recommended. It is a good indicator that the raw water particle charge has been neutralized. However, it is recommended that the SCM only be used for monitoring purposes and not for direct control of the chemical feed systems.

An SCM is a valuable tool. However, it must be calibrated frequently to take into account seasonal water variations as a minimum. The SCM set point may also change depending on the raw water quality.

Example. One plant used an SCM for control purposes, and it did not correctly interpret the changing raw water conditions in one instance. As a result, the plant dramatically overfed alum, thereby upsetting the entire system. The entire contents of the plant had to be discharged to waste resulting in a large waste of water, time, and chemicals.

Laboratory Sample Water

A sample of chemically treated water is also included, after the chemical feed and flash mixer, and then directed to the laboratory for the operators' use. It can be used by the operators for verification and optimization of the chemical feed process.

STAGE 6—OPTIONAL CLARIFICATION MONITORING

In a conventional treatment plant as shown in Figure 8-9, monitoring of the clarified effluent may be desirable. In some cases, the regulatory agencies may require that a turbidimeter be installed at this location to ensure that the treatability standards are being met (Figure 8-9). Membrane treatment plants may require an SDI at this location to ensure that treatability standards are being met prior to membrane filtration.

NOTE: As stated frequently, chemical feed systems and instrumentation must be tailored to the specific site and the treatment processes being used.

STAGE 7—FILTER EFFLUENT INSTRUMENTATION

The instrumentation shown in Figure 8-10 might be provided for each filter in a conventional plant or perhaps for a bank or module of membrane filtration units. The instrumentation shown is also consistent with that which might be provided for a filter with a constant effluent flow control mode of operation, as discussed in the next section. A level control device is also included in the clearwell, to be used for final effluent pump control.

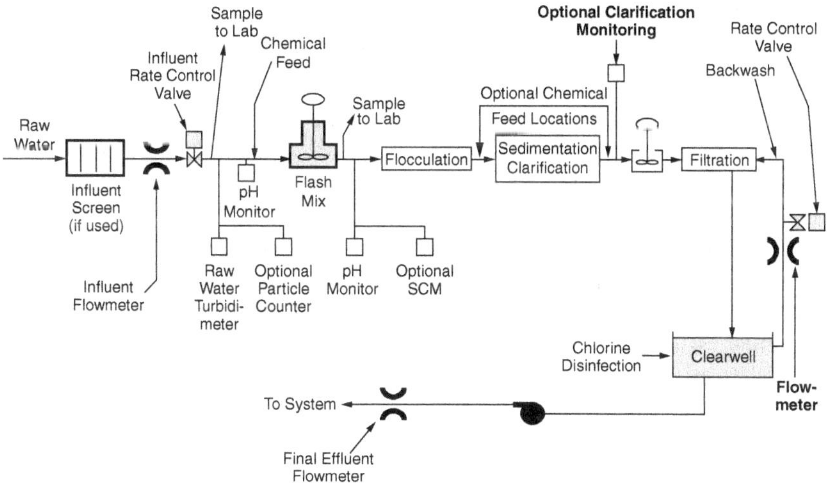

Figure 8-9 Process flow diagram, Stage 6: Optional clarification monitoring

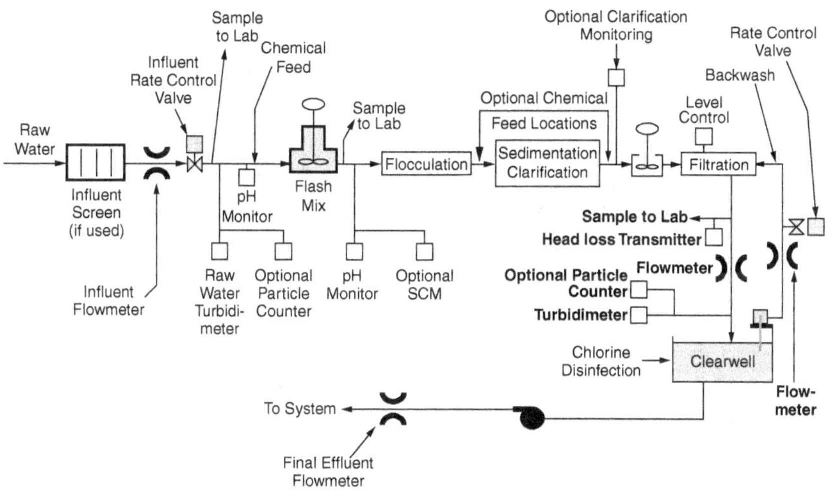

Figure 8-10 Process flow diagram, Stage 7: Filter effluent instrumentation

Filter Instrumentation

A number of different types of instrumentation are required for proper operation of a filter, depending on the type of control methods used. The instrumentation described below is for constant rate operation.

Head loss transmitter. A head loss transmitter is used for initiating backwash and for monitoring the rate of head loss development. The use of a simple on/off switch is not recommended.

Individual filter effluent flowmeter. An individual meter is used for modulating the rate control valve for each filter in order to maintain constant effluent flow.

Turbidimeters. Turbidimeters are used for monitoring effluent water quality and to ensure that each filter is meeting the desired standards. Separate turbidimeters for each filter can also be used to identify any filter that may not be operating as well as the others.

Optional particle counter. Although it is recommended, not all individual filters have their own particle counter. However, each filter will probably be required to have one in the future. A particle counter is a very valuable tool for monitoring the efficiency of a filter, which will be discussed in greater detail later on in this handbook.

NOTE: The use of both turbidimeters and particle counters on each filter also provides an early warning of failures and will greatly assist the operators in troubleshooting.

Clearwell Level Control

A level controller in the clearwell may serve two functions: (1) to ensure that adequate water is available in the clearwell for backwash purposes, and (2) to control the operation of the final distribution pumps.

In the case of a two-compartment clearwell (one for CT purposes, plus a pump well), the backwash supply should come from the CT compartment, while the final effluent pumping should come from the pump well. Two level controllers might then be required, one for each of the two clearwell compartments.

Laboratory Sample Water

Another laboratory sample pipeline is included in Figure 8-10, which comes off the filter effluent piping. Again, this sample pipeline is for the operator's convenience and may be used to calibrate and verify the operation of the instrumentation. If there are numerous filters, there may be too many to pipe them all to the laboratory individually. In that case, the designer should coordinate with the owner/operators to provide a representative number of samples.

STAGE 8—FINAL pH ADJUSTMENT

Final pH adjustment is included on the clearwell discharge in this stage (Figure 8-11). For discussion purposes, it is assumed that either lime or soda ash will be used to raise the pH after coagulation and filtration (if required). The reason it is shown on the clearwell discharge in Figure 8-11 is that chlorine disinfection for CT purposes is more

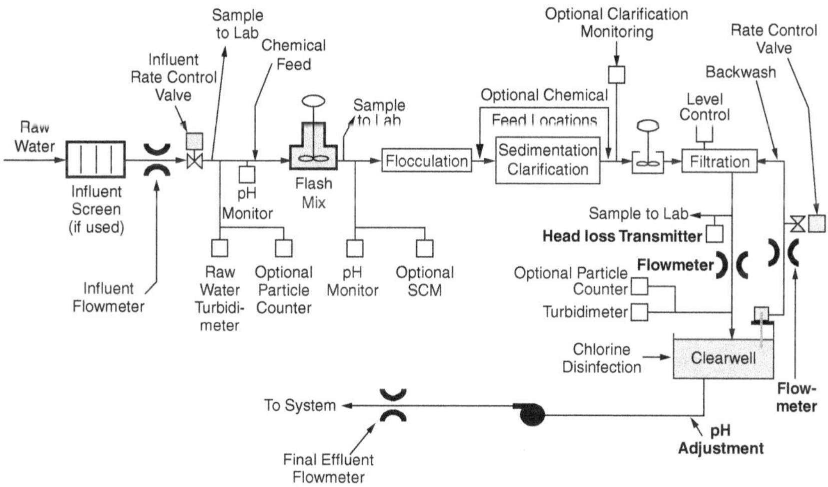

Figure 8-11 Process flow diagram, Stage 8: Final pH adjustment

efficient at a low pH. Actually, the pH buffering chemical might even be added after the final distribution pumps for mixing purposes.

However the pH adjustment chemical is fed into the process water, adequate mixing is required to ensure that the final instrumentation detects a representative water sample for an accurate reading. For example, undissolved chemicals can result in artificially high turbidity and particle count readings in the final instrumentation.

STAGE 9—FINAL EFFLUENT INSTRUMENTATION

The final effluent instrumentation is intended to measure the actual water quality being delivered into the system (Figure 8-12).

Chlorine Residual Analyzer

The primary function of this analyzer is to measure the actual chlorine concentration delivered to the system. Alarms are usually included in case of dosages that are too high or too low. One of the other purposes of this analyzer may be to control the amount of chlorine added in the clearwell, to produce the correct concentration. If that is the case, a time delay should be built into the controls to make sure that changes have had adequate time to equalize in the process water before any further changes are made.

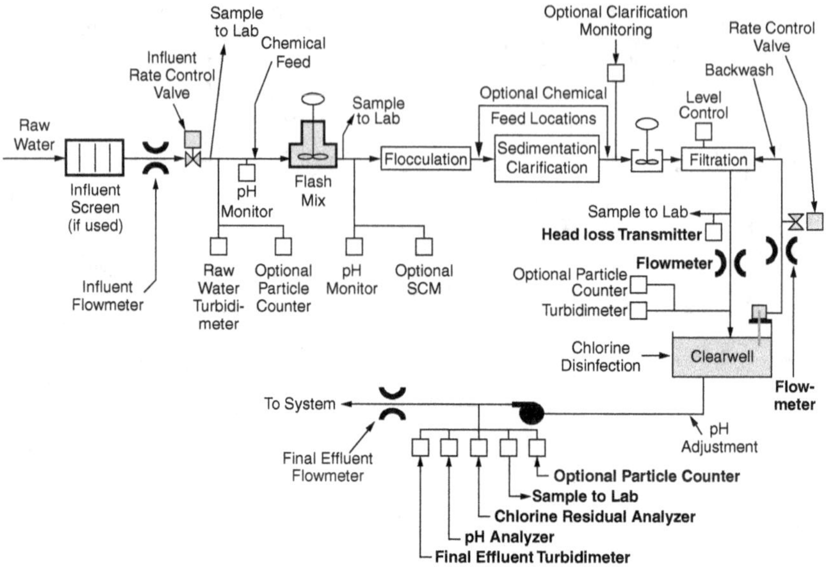

Figure 8-12 Process flow diagram, Stage 9: Final effluent instrumentation

pH Monitor/Analyzer

The pH monitor on the plant effluent may be used to control and/or to verify that the proper amount of pH adjustment chemical has been added after the clearwell. The operator establishes a desired set point and enters it in the plant controls or on the meter itself, depending on how the controls are designed. On the one hand, if the actual measurement is too low, a signal is sent to the chemical feed system and more buffering chemical is added. On the other hand, if the reading is too high, the chemical is reduced. In either case, there will be a time delay between changing the dosage and having the new amount show up on the pH monitor. Therefore, a time delay must be built into the controls so that multiple changes will not be made until there has been adequate time for the changes to reach and be read by the pH monitor. Otherwise, there could be a problem either underfeeding or overfeeding the chemical. If the readings are too high or too low, an alarm can be sounded to notify the operator of the problem.

Final Effluent Turbidity Monitor/Particle Counter

These two instruments are intended to monitor the final effluent water quality delivered to the system and to ensure that the regulatory standards are being met. Low readings, below the set points, are to be

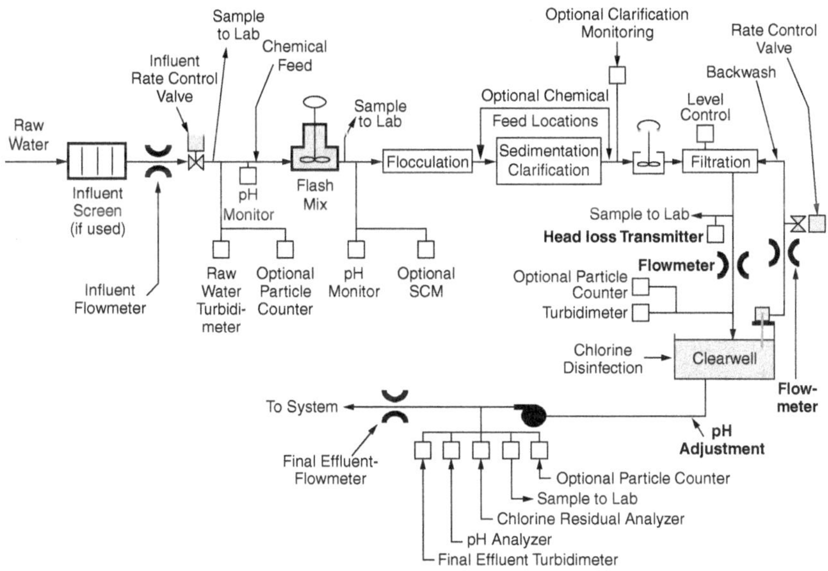

Figure 8-13 Completed process flow diagram

desired on both instruments. Rising particle counts or turbidity may mean that the filters need to be backwashed. Rising particle counts may indicate that a backwash is desired before the turbidity indicates a backwash is desired. Very high readings on either instrument may sound an alarm that tells the operator that there may be a serious problem to be identified and resolved.

Laboratory Sample Water

A final water sample should be piped to the laboratory from this location for the operators use in calibrating instruments and to verify the final effluent quality. Although the number of water samples from the filters may be limited, the final effluent sample should be provided regardless.

The Completed Process Flow Diagram

The completed process flow diagram is shown in Figure 8-13 and should be used to develop an operational process control philosophy and manual. Although the diagram illustrates the general location of instrumentation, it is not to scale and only illustrates the sequence of installation. It should also be used as the blueprint for the design of the plant operating equipment.

It should be noted again that this is a generalized type of flow diagram. For example, only flow control valves are shown and only one filter. A completed flow diagram might show all operating valves and other equipment. In addition, multiple filters may be shown, or one may be shown and labeled as typical of many. Also, there may be multiple pumps where only one is shown in Figure 8-13 at each location.

SUMMARY

All the process components have now been located on the flow diagram in their proper places, relative to each other. It is highly recommended that the plant operators be very familiar with this flow diagram, as it illustrates the relation of all the equipment and instrumentation components to each other. The next step is to develop a means of controlling all these devices.

CHAPTER NINE

Treatment Plant Controls

The next step in the design process is to develop an operational control philosophy as a framework for the control system, eventually consisting of both hardware and software. Implementation of this step is extremely important to the proper operation of the plant, and as such needs input from designers who have operational knowledge. Its importance is such that information from a variety of separate equipment suppliers cannot alone be relied on. Operational experience by the controls designer is then of extreme importance.

NOTE: These comments are provided to assist the operators in understanding how the various plant components are supposed to work and then obtaining and understanding accurate data with which to operate the plant.

FILTER CONTROL MODES

The first control strategy to be considered is that required by the filters (granular media or membrane). The rest of the plant is then designed according to the needs of these filters. In the case of granular media filters used for pretreatment for membrane filters, a separate strategy for controlling each would be required.

A number of common filter control strategies should be considered, including the following:

- **Constant Effluent Rate**—Each filter operates on the same constant effluent rate, as established by the operator.
- **Declining Rate**—A common supply pipe (pressure) or influent channel (gravity) to all filters is used, with no effluent rate control. The cleanest filter takes the highest flow.
- **Equal Loading/Constant Level (Gravity Operation Only)**—The flow is split equally to all filters, and the effluent flow is the same as the influent.

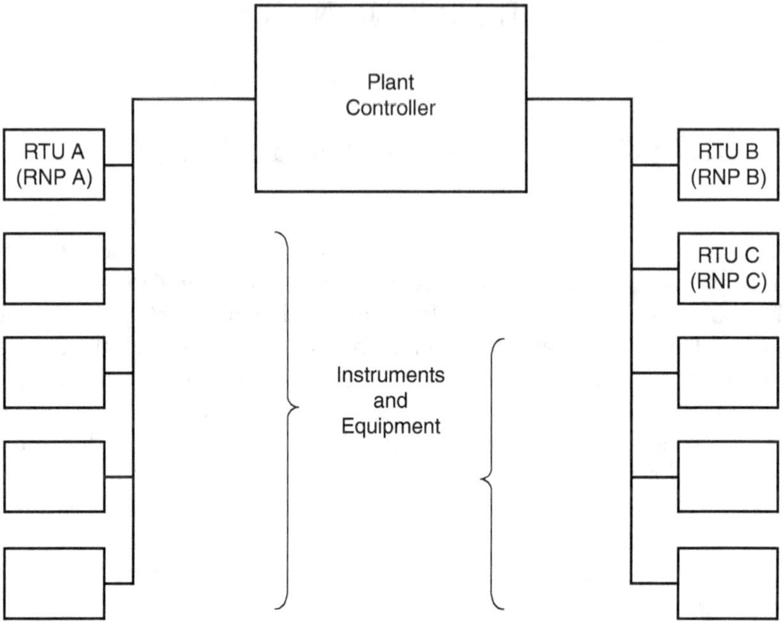

NOTES:

1. Only instruments and equipment that cannot communicate directly using the chosen fieldbus protocol are connected to remote panels.
2. Communications are "Report by Exception" with a background "heartbeat" to ensure that a connection is maintained.
3. Results in faster overall communication.

Figure 9-1 Plant communication diagram using a Device Net type protocol

- **Equal Loading/Variable Level (Also Gravity)**—Equal loading is achieved by having the water flow over weirs into each filter. The effluent from the filters also flows over weirs that set the starting level of water over the filters. As head loss builds up, the water level rises to create the necessary pressure to maintain the flow rate set by the incoming weirs. Refer also to Reference 1 for an additional discussion of filter control types.

For the purposes of this discussion, a constant rate mode of operation will be assumed, which means that each filter will be operated at the same flow as all the others. A diagram of the plant communications (between the various instruments and the plant controller) for a typical constant rate granular media filter is shown in Figure 9-1. It should be noted that a Device Net type protocol is assumed in this figure,

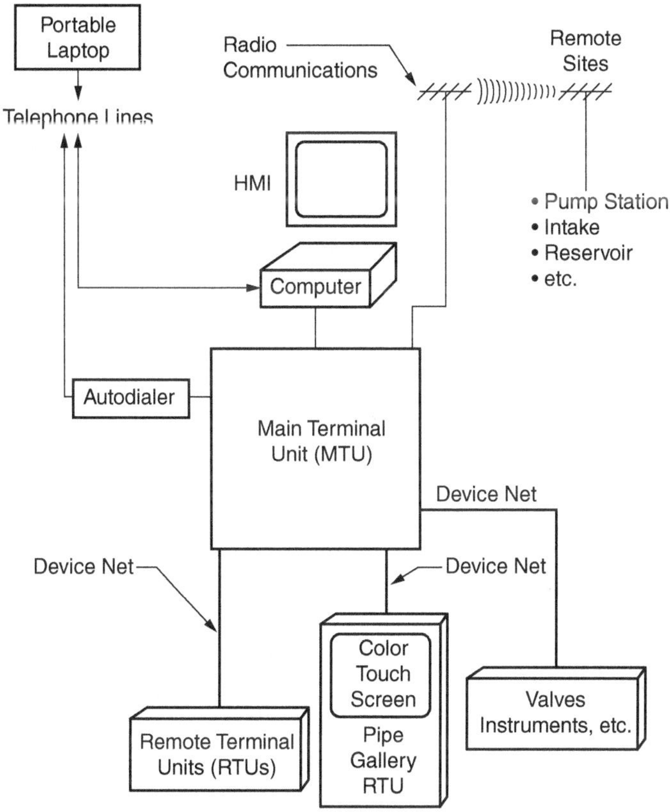

Figure 9-2 Plant communications including remote systems

which means that the communication for a group of instrumentation or equipment components travels on one wire. The manufacturer should be contacted for a more complete description. Another diagram of the same plant is shown in Figure 9-2, which also includes the remote communication systems.

Note: A more complete discussion of filter control types is included in Reference 1.

INDIVIDUAL FILTER OPERATION

Operating Water Level

When there are multiple filters, the operating water level is often intended to be the same for each one. Maintaining the proper operating water level requires a separate set of controls and instrumentation for that purpose. For example, if the overall plant flow rate is constant, the operating water level may change when one or more filters are off-line for backwash or maintenance purposes. It may also be possible to index the rate up for the remaining filters, to account for the difference, which would have the effect of maintaining the same influent level.

Example

One treatment plant had a large number of filters, which were fed by a very long common influent channel. Due to the length of the influent channel, the water level actually varied from one end to the other. As a result, the operating water level also varied slightly from one filter to another, which complicated the overall filter control. The intent was to resolve this problem with a new control system that was then under construction.

Constant Rate Control

As shown in Figure 9-3, the constant rate controls consist of a flow-meter and effluent flow control valve for each filter. A control loop (software) modulates the rate control valve to maintain a set flow rate. As head loss builds, the rate control valve opens slightly to compensate.

NOTE: The flow rate set point should be screen adjustable, for each separate filter, by the operator.

Increasing Head Loss

As head loss increases, the effluent flow would normally begin to decrease for a given filter. The flow control valve for that filter should then modulate slightly further open to reduce back pressure and thereby maintain a constant flow rate.

When the effluent flow control valve opens to a set maximum percentage (70 or 80 percent, for example), the constant rate cannot be maintained further by the valve, and the filter should be backwashed.

Multiple Filters

The control of individual filters, as previously described, is relatively simple. More complex controls are needed with multiple filters, however.

For discussion purposes, a treatment plant with 10 filters will be assumed. The steady-state operation for each filter will be as previously described, with an operational level control and separate filter flow controls. Then, when one filter goes into backwash, the filtration capacity is decreased by 10 percent, and some changes are required.

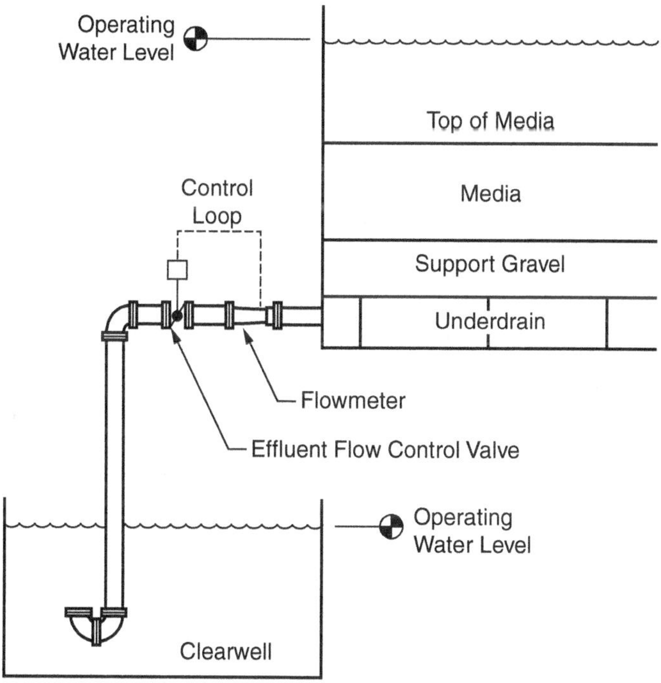

NOTES:

1. As headloss increases, the flow control valve modulates to maintain a constant flow rate.
2. When the effluent flow control valve reaches 70–80% open, the filter must be backwashed.

Figure 9-3 Constant effluent rate instrumentation

Several options for controlling flow are discussed in the following sections.

Reduce the plant flow rate. Reductions in flow rate can be accomplished by turning a pump off or by throttling back the plant flow with an influent rate control valve. Turning a pump off would require that a number of influent pumps be provided, at least one of which would have a 10 percent capacity. Implementing this option is cumbersome and expensive and is seldom implemented.

Throttling with a flow control valve. Using a flow control valve to throttle the plant influent flow is relatively simple and is often practiced. Implementing this method does require that the pumps have adequate capability for this purpose, because they will back up on their curve when the valve is throttled back.

NOTE: Throttling valves must be sized properly for the purpose. Generally, they are at least a size smaller than the process piping, especially for larger sizes of pipe. It is recommended that the valves should be sized so that they modulate between 30 and 60 percent when open. The valve manufacturer should be contacted for recommendations on sizing.

Variable speed pumps. The use of variable speed pumps is also very common for reducing the plant flow rate. Using this method requires that the pumps be capable of producing the necessary pressure at the reduced flow condition. Then, when back to full capacity, the pumps will produce excess pressure, which must also be considered.

NOTE: Reference 2 contains an extended discussion of variable speed pumps.

Increased unit flow rate. When one filter is out of service or in backwash, another alternative for handling the plant flow is to increase the unit rate to each of the filters remaining in operation. With one filter out, the others would then be required to operate at a unit rate of about 10 percent higher than normal. In order to accomplish this, all the flow set points for the individual filters would have to be temporarily reset by the controls for this time period.

Implementing this option would require that the filters be operated normally at a rate at least 10 percent lower than design. Then, at the increased flow condition they would still be at, or less than, the design rate.

A potential hazard of operating in this mode is that the filters would all experience a rapid increase in flow rate, which may produce poorer water quality during that time. Here again, it is recommended that this condition be tested during the pilot study program to determine if acceptable water quality could be produced at the higher rates.

Standby filter(s). The recommended control option is to have a spare filter off-line, which can be brought on line when another filter is taken off-line for backwash. Having a spare filter obviously adds to the plant cost. However, it has considerable value. It can also be used in the event that a filter has to be taken off-line for inspection or maintenance. Being able to help maintain the design plant flow in case of an emergency is of significant value in and of itself.

In order to implement this option, it is necessary to carefully consider the exact timing and sequence of events, and especially valve speed. It is very difficult to take one filter off-line and add another, without some small spikes in the flow rate to the individual filters.

These issues will be discussed in more detail in the analysis of operational problems included in this handbook.

PUMP AND VALVE CONTROLS

Once the filter controls are completed, the next step is to establish the influent controls. The influent pump(s) must operate to match the filter control mode of operation previously established. The controls may use variable speed pumps (VSP), multiple pumps, or a throttling valve.

Spare Pumps (Process or Chemical Pumps)

A full size spare pump is recommended for each application, so that flow can be maintained in case of failure. In the case of multiple process pumps, the spare need be only as big as the others. A lead/lag type relationship should be established for each pump set, wherein the lead pump comes on first with the spare, or lag pump, in standby in case of failure.

With multiple pumps, including a spare, the operation of all of them can be rotated to ensure even wear. The rotation usually occurs when one is started up, which allows another to be turned off. In case of a failure, the standby or lag pump should be activated.

Chemical Pump Control

The chemical pumps are commonly flow paced proportional to the raw water flow rate, as discussed in this handbook. The pumps must be provided with variable speed control for this purpose. Duplex pumps are also recommended for each chemical, as previously discussed.

TYPICAL LOGIC

A number of different logic types are used for control in treatment plants, including pixel graphics and "object block" graphics, which are recommended.

Graphics Program

A computerized control system using a graphics program that provides illustrations of all the operating components is recommended. On typical screens, levels are shown both digitally and on a sliding scale. Components change color as they are activated, all timers in the program are screen adjustable and are shown on the screen, and all push buttons and selector switches are shown, which can be activated by clicking on them with the computer mouse.

Plant Control

Using the type of computer control discussed herein, the plant can be started or stopped, valves can be opened or closed, and all process pumps and chemical pumps can be started or stopped automatically or manually with the computer mouse. In addition, history screens should be available for all analog or variable data.

Control Capability

The plant operator should also be able to operate the plant either manually or automatically, change chemical feed settings, and change the time for any required activity. In other words, the operator should be able to modify everything except the actual sequence of events in the control logic, including the backwash cycle, level and flow control, and so on.

A higher level of security could be provided for logic modifications and should be available only for the system programmer, the plant manager, or the senior operator as required for that particular treatment plant.

Control Logic

When a plant is being designed, the controls and the plant's operating philosophy are often described in writing. However, the recommended method is to use logic diagrams that use decision diamonds, action blocks, and that also show all desired pushbuttons, switches, alarms, and timers. A portion of one such a control diagram is shown on Figure 9-4. Some text describing the operational philosophy may also be used.

In either case, a very detailed description of the plant operation should be included. If a logic diagram is available, it may make future troubleshooting much simpler. It may also be simpler and easier for the plant operators to understand than numerous sheets of ladder logic.

Example

At one treatment plant, the designer did not provide any direction for the controls programming in the original construction specifications. The programmer was forced to use input from a number of different equipment suppliers. The result was uncoordinated and caused damage in the filters. Someone needs to be in charge of the development of the plant controls who has significant process experience as well as familiarity with the equipment.

Proportional Integral Derivative (PID) Control

There are numerous situations in a treatment plant where a variable signal, such as from a flowmeter or level transmitter, is used to

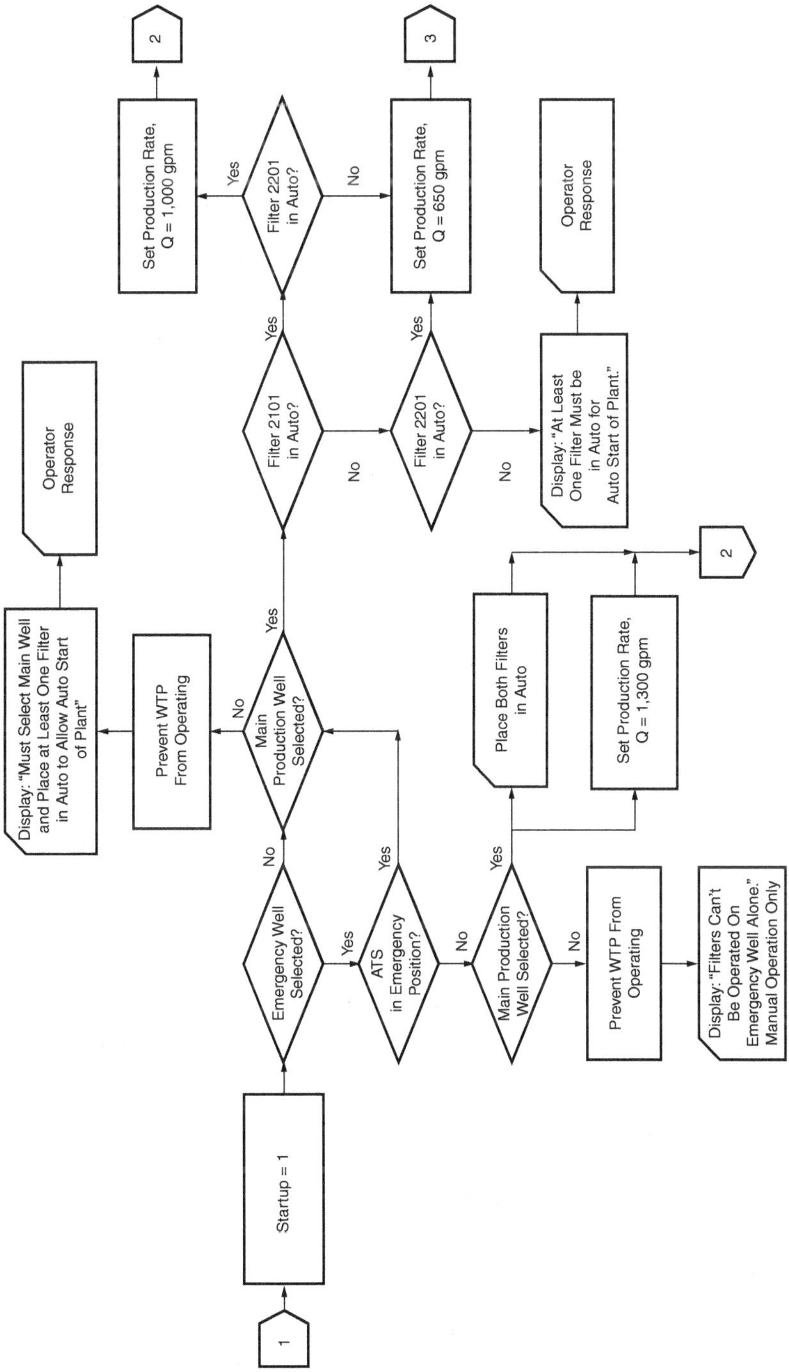

Figure 9-4 Typical plant logic

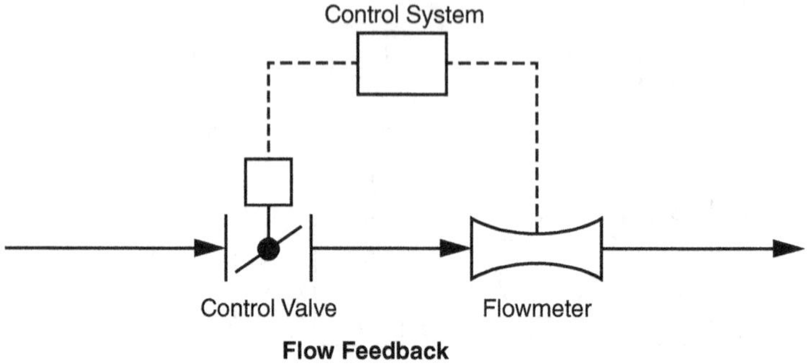

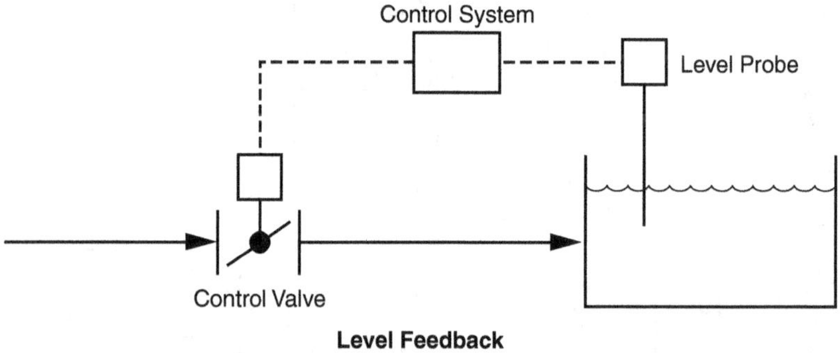

NOTE: A feedback control method typically based on flow or level.

Figure 9-5 Proportional integral derivative (PID) control

modulate a valve or pump. An example is for maintaining constant effluent flow rate, as discussed previously. Typical process control diagrams of these loops are shown in Figure 9-5. These systems usually operate according to a PID control system as shown in Figure 9-6.

PID factors. A PID control system has several variables including proportional gain, integral factor, derivative factor, and deadband. These factors determine the magnitude and speed of a response to a change in conditions. The figures illustrate these factors.

Deadband. It is nearly impossible for a valve to achieve the exact position required to produce the precise flow required. The deadband is a window or range in which the conditions (flow, level, etc.) are assumed to be met. For example, the valve may be moved until the flow is ±50 gpm of the set point. Because of the importance of a deadband, it is recommended that they all be screen adjustable for the operators' use.

Screen adjustable. The reason for discussing PID control is that it is recommended that the variables previously described should be screen adjustable and that the plant operators should be trained in

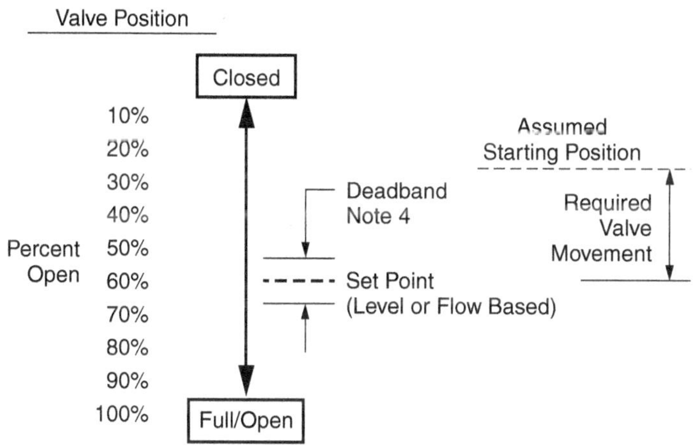

NOTES:

1. Proportional Gain — Changes magnitude of output signal.
2. Integral Factor — Adjusts output depending on magnitude of difference between actual flow or level value and desired.
3. Derivative Factor — Adjusts the output to the valve based on the rate of change of the actual process value compared to the set point.
4. Deadband — An range in which the set point is assumed to be met.

Figure 9-6 Proportional integral derivative control system.

their use and application. Accurate control of the plant components is extremely important. There have been numerous plant control systems where the programmers had neither the knowledge nor the training to put the system together. Therefore, the plant operators may need to have the capability to fine tune the system during or after startup.

SUMMARY

A brief discussion of some of the types of plant controls available is included in the previous sections. It should be noted that these controls are site specific to each separate plant control system. The operators need to be intimately familiar with the controls at their plant and how the system functions. Such familiarity is required in order to control the plant properly, especially when operating the plant manually or under changing conditions.

Developing the necessary hardware for control is discussed in the following chapter.

CHAPTER TEN

Computer Control Hardware

Once the plant operational control philosophy has been completed, the next step is the identification of the control hardware needed for implementation (Figure 10-1). As mentioned previously, this handbook assumes the use of an up-to-date computer control system. Such a system may include a number of components as described in the following sections.

PROGRAMMABLE LOGIC CONTROLLER (PLC)

The "heart" of a computer control system is, of course, the computer or PLC. Ordinarily, the operational staff may have little choice in the type of PLC to be used, unless it is intended to match other existing equipment. It is recommended that the PLC be of industrial grade, with parts and technical support located reasonably close to the plant site.

The graphics software should be of the "object block" type, which is widely available commercially, with knowledgeable technical support also located close by. Local technical support for both the PLC and the software used is highly important.

MAIN TERMINAL UNIT (MTU)/REMOTE TERMINAL UNIT (RTU)

Most control systems of the type discussed in this handbook have an MTU and one or more RTUs. Refer to Figure 10-1 for a graphic representation of the MTU and RTUs.

MTU

The MTU is an electrical panel that typically contains the PLC and all the wiring connections to and from all RTUs and other plant components. The size of the panel varies widely depending on the size of the plant, the type of communications system used such as Device Net, and whether or not selector switches, push buttons, and indicating lights are included.

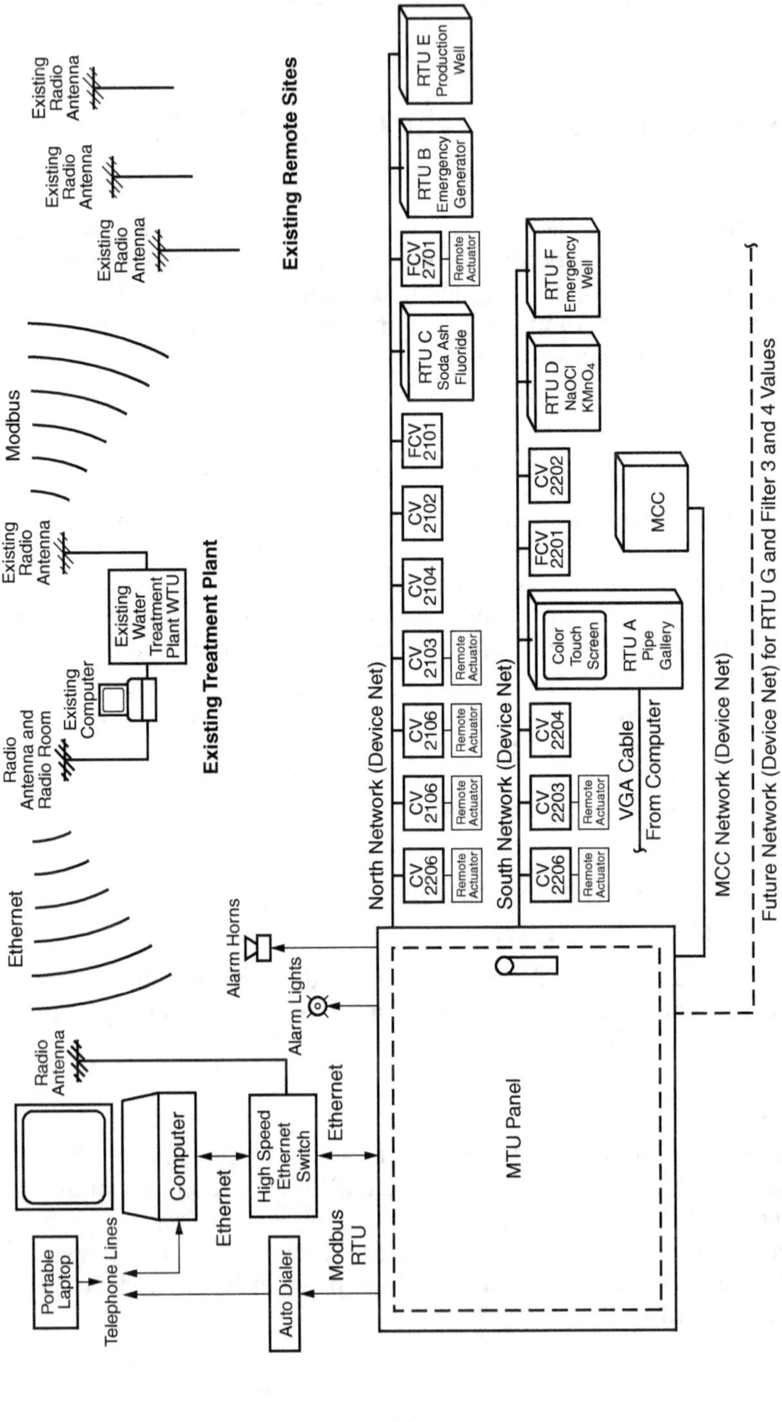

Figure 10-1 Telemetry and control flow diagram

RTU

An RTU usually consists of a small PLC in a panel remote from the MTU, which is used to collect data from instrumentation and to communicate between the various operating components of the plant and the MTU. Each RTU is then connected to the MTU separately or by a common communication system such as Device Net.

If all instrumentation and operating components are wired to the control system separately, a large number of RTUs may be required. However, if a computer communications system such as Device Net is used, there may be fewer RTUs, depending on the communication capability of the individual instrumentation devices and other operating components.

HUMAN MACHINE INTERFACE (HMI)

The HMI may consist of a benchtop computer with a monitor that is used to display the graphics control program. It is also connected to the MTU as shown on Figure 10-1. The operator may select from a variety of screens to monitor and control the plant.

At this level of design, input from the operational staff is extremely valuable in determining how they wish to operate the plant. For example, the operators may want all the controls concentrated in one room (the control room). Or, they may also wish to have a number of local or satellite control stations in different areas of the plant.

A small plant may only need a single control location. However, a large plant, or one with several floors in the building or multiple buildings, may require control stations in a number of different locations. Here again, the operators should decide what they need to operate the plant.

TOUCH SCREEN CONTROL

Touch screen control may also be desired by the operational staff, especially in remote locations where there may only be a wall mounted control cabinet or RTU. Using this type of system, the operator need only touch the monitor screen on the function desired, with no mouse required. It is recommended that the operators be given the opportunity to select the location and type of control stations they need.

COMMUNICATION TYPES

There are a number of different types of communications commonly used in treatment plants: communications between the main plant and remote locations such as pump stations and reservoirs, and computer communications between the MTU/PLC and the local plant instrumentation and operating components.

Remote Communications/Telemetry

Remote telemetry typically uses either radios or telephone communications. The choice is usually made by the owner/operator depending on the reliability of either type. In some places, radio communications may not be feasible. However, telephone communications may not have the desired level of reliability in some other cases, which is required by the owner/operator. Both are common.

Computer Communications

The type of computer communication used can have a dramatic impact on plant operations, especially considering the "response time" that results.

For the purposes of this discussion, a Device Net or Ethernet type of communication will be used and is recommended. Figure 10-1 illustrates a typical plant control system using Device Net communications. Both types will be described in more detail in following chapters.

Control Response Time

Control response time is defined as the elapsed time from when an activity is initiated by the computer until that activity actually occurs. For example, the time required for a valve to actuate once the computer has sent the signal or the operator has initiated it on the screen.

NOTE: Another type of response time is the time required for a change in water quality to be recognized by the local instrument and for that data to be transmitted to the computer. Both types are important to the operation of the plant.

SUMMARY

The process controls represent the most common cause of operational problems. Therefore, extreme care should be exercised in providing a workable control system that is easy to understand and operate. For best results, it is recommended that the control system designer have significant process operational experience, and that the plant operators be consulted and then be highly trained in its use. The intent is to provide the operators with the utmost flexibility in operating the plant.

NOTES: Response time for both controls and process is an important part of plant operations and will be discussed in more detail later

Table 10-1 Brief controls troubleshooting guide

Problem	Probable Cause	Proposed Solutions
Excessive valve hunting and seeking	• Narrow deadband • Proportional control out of adjustment • Rate control valves too large	• Change deadband • Modify proportional controls • Provide a smaller valve
Uncontrolled air	• Failed valve(s) • Improper controls sequence	• Replace/rehabilitate valves • Modify control to purge air
Long response time	• Programming issue	• Evaluate programming: In some cases, resolving this issue may require extensive reprogramming • Evaluate system components
Hydraulic shock	• Failed valve(s) • Improper controls sequence • Actuators too fast	• Replace/rehabilitate valves • Check controls sequence • Replace actuators

on in this text. A very brief troubleshooting guide for controls is contained in Table 10-1. For a complete filter troubleshooting guide refer to Reference 1.

Example

At one plant, the operators had to look up a particular control step in a book and then enter the proper code into the computer to determine what was actually taking place for that one step. In another plant, the designer allowed a variety of manufacturers to independently influence the design of the controls, which resulted in an unworkable system and a major lawsuit.

CHAPTER ELEVEN

Instrument Installation and Chemical Locations

The proper location, installation, and orientation of instruments can have a major impact on producing accurate information and on the evaluation of that information. The following discussion is included to provide the operators with information on troubleshooting instrumentation and information on determining the proper installation methods to be used.

NOTE: The comments in this chapter are also provided to assist the operators in obtaining accurate data from the instrumentation.

TRANSMITTER LOCATION FOR VENTURI FLOWMETERS

Although Venturi flowmeters may not be used as much as formerly, the pressure transmitter location for them is critical. As shown in Figure 11-1, the pressure transmitter must be located below the Venturi tube to prevent air from entering the tubing. Because air is compressible, the wrong results will often be indicated when air is present. Locating the transmitter below the Venturi tube allows air to be bled upward and out, without influencing the flow readings.

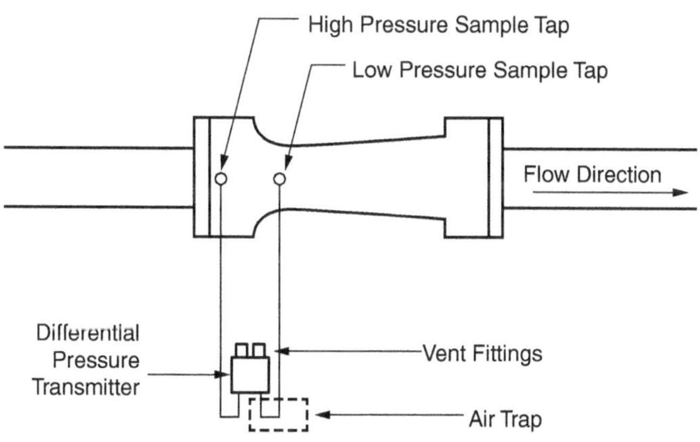

Figure 11-1 Transmitter location for Venturi or orifice flowmeters

Example

The Venturi flowmeters at one treatment plant did not work and had not ever worked properly because the pressure transmitters were located above the pipeline and were all filled with air. A recommendation was made to the operators to relocate the pressure transmitters below the sample point to prevent air from entering.

LOCATIONS FOR OTHER FLOWMETER TYPES

For other flowmeter types, the location of the meter itself can be very important. For example, it is recommended that turbulence and air pockets (high loops in piping) should be avoided. Some manufacturers may say their equipment is not affected by those factors. However, unless the designer has specific experience with a particular type of flowmeter, it is recommended that a conservative approach be taken. The flowmeters should be located so that they have an amount of straight run of process piping between pipe fittings that might cause turbulence. In addition, try to locate them in a low area where there could be no accumulation of air, even if such an area has to be constructed especially for the purpose.

Turbulence

It is recommended that the flowmeter not be installed close to pipe fittings or components that might cause turbulence. Some amount of straight piping in front and behind the flowmeter is recommended for best results. The amount of straight pipe required for a particular flowmeter may vary. The manufacturer should be contacted for recommendations.

Partially Filled Pipe

Again, some manufacturers state that the accuracy of their flowmeters is not affected by piping that is only partially filled with water. If the operators have confidence in the accuracy of their flowmeters, it is all well and good. However, if there are irregular or erratic results, an investigation may be necessary to determine if there is an accumulation of air in the flowmeter and piping.

Example

At one installation, a buried pipeline could drain under some circumstances, which then caused erratic results from the flowmeter. It was necessary to install an elbow downstream of the flowmeter to keep the pipeline full of water.

OTHER TRANSMITTER LOCATIONS

All other pressure transmitters should be located in a similar manner to the Venturi flow units; that is, the transmitters should be located below any process connections that are subject to air contamination.

PROCESS CONNECTIONS FOR INSTRUMENTATION

Sample connections for instrumentation should always come off the side of the process piping. Connections on the top are prone to collecting air, while connections on the bottom may collect accumulated slimes.

Example

A sample connection was installed on the top of a large pipeline at one plant. It was originally designed to use an injection quill to allow the sample to be drawn from the center of the pipe. However, the injection quill was broken during installation and not replaced or repaired. As a result, inaccurate results were obtained, and the connection had to be reconstructed.

CHEMICAL FEED CONNECTIONS

Although not directly related to this discussion, schematics for recommended chemical feed connections are shown in Figure 11-2 for the operators benefit. The purpose for including them is to help reduce operational failures and improve reliability.

TURBIDIMETER SAMPLE

A schematic of a turbidimeter installation on filter effluent piping is shown in Figure 11-3. Two items are of interest in this figure; the sample tap location and the use of a sample pump.

Sample Tap Location

As a general rule, sample taps for any purpose should be installed on the side of the pipe for more accurate results, as previously discussed. Taps on top of the pipe are prone to picking up air bubbles. Taps on the bottom of the pipe may collect sediment or slimes that may accumulate in the piping over time.

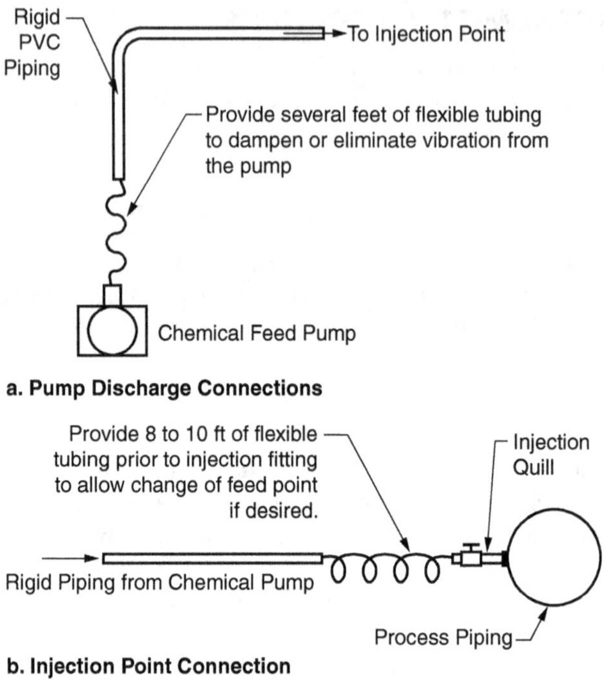

Figure 11-2 Chemical feed piping connections

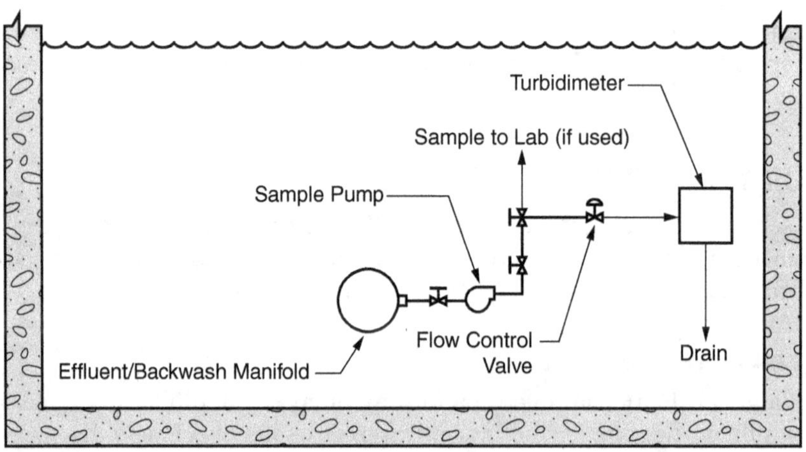

NOTES:

1. Sample tap should be installed on the side of the effluent manifold.
2. A pump is required for filter effluent sampling because of negative pressure at the end of the run.

Figure 11-3 Filter effluent sample

Turbidimeter Sample Pump

The pressure in the effluent of a filter is subject to change. In many gravity filters, the pressure may even be negative at the end of a filter run. In that case, water in the sample piping may be pulled back into the process piping, which then allows air to enter the turbidimeter body resulting in erroneous readings. Therefore, a small sample pump is always recommended, for all sample locations, to eliminate this problem.

NOTES:

- The flow in the process piping should be maintained by a water seal in the discharge to the clearwell located below this level.
- If a negative pressure occurs or is possible, it will be necessary to provide a sample pump to maintain a supply of sample water to the turbidimeter. They are usually required in gravity filters.
- Negative pressures are common in the effluent of gravity filters, especially if the driving head over the top of the filter is less than 6 ft and if the terminal head loss is 8 ft or more. If there is any question, a low range pressure/vacuum gauge installed at this location may be helpful. Please also refer to Reference 1 for additional information.

FINAL EFFLUENT/CLEARWELL TURBIDITY SAMPLE

A turbidity sample on or near the discharge of a vertical line shaft turbine pump, as shown in Figure 11-4, is potentially subject to considerable problems with air when the pump starts. When the pump is off, the water level in the pump column nearly always drains down to the level of the clearwell. Then, when the pump starts, a large volume of air is delivered to the piping immediately, even when an air release valve is present as shown in this figure.

Air Trap

To help prevent air from entering the sample tubing, an air trap can be created by installing a low loop as shown and is recommended in all such cases.

External Bubble Trap

Turbidimeters may have their own internal bubble trap. However, when large amounts of air may be present, an external bubble trap is recommended as shown. For a conservative design, the use of an external bubble trap is recommended on all installations where air may be

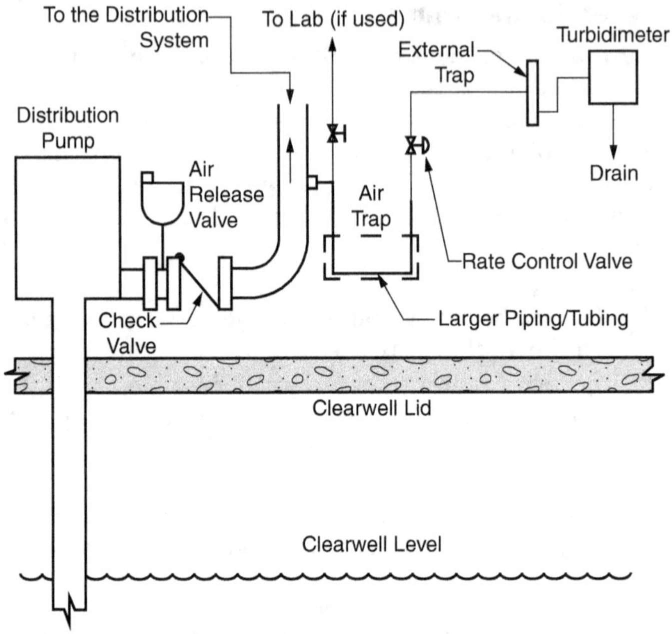

NOTES:

1. An air trap should be provided before effluent instrumentation as shown. An external air trap is also recommended in front of the instruments.
2. A size larger piping/tubing should be used for the air trap than is required for the instruments.

Figure 11-4 Final effluent sample

present. The manufacturer should be contacted for an accessory for this purpose or one may be constructed separately.

HEAD LOSS TRANSMITTER LOCATION

A head loss transmitter is recommended and is normally provided on the effluent of each filter, as shown in Figure 11-5. It should be noted that such a transmitter will measure the pressure at the elevation at which it is installed. The desired elevation for installation of the transmitter may vary according to the amount of negative pressure anticipated. For example, if a negative pressure is anticipated in the effluent piping, it may be desirable for the transmitter pressure to stay

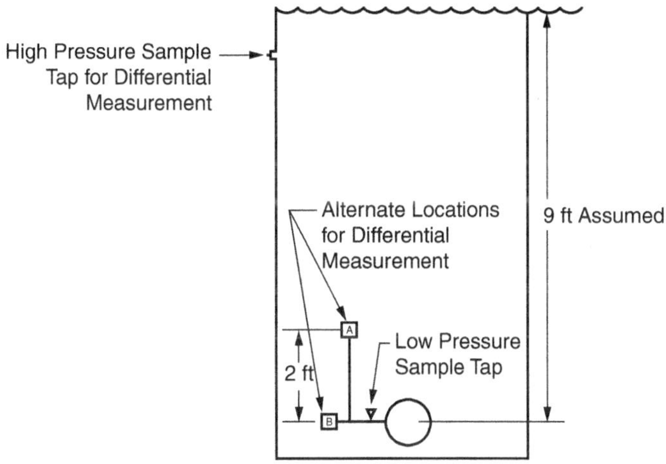

Assumptions:

- Assume 3 ft clean headloss for media and underdrain.
- Assume 8 ft operational headloss.
- Assume 11 ft of total driving head.

Pressure Range:

Location A—Pressure range +7 to –4 ft.
Location B—Pressure range +9 to –2 ft.

NOTES:

1. The differential is the same although the calibration of the transmitter is different.
2. If the water level varies, a straight pressure switch will indicate a lower value. A differential switch or transmitter will maintain a similar value, although the pressure in the effluent/backwash manifold will be more negative for Location A.

Figure 11-5 Head loss/switch transmitter location

above zero, depending on the type of transmitter used. For a differential pressure transmitter, the height may not be critical.

If a straight pressure-only transmitter is used, the height may be critical. To maintain a pressure above zero, the transmitter could be raised slightly more than the anticipated amount of negative pressure. For example, if the negative pressure is not expected to exceed 2 ft of water, a pressure transmitter could be located at a height of 2 ft above the effluent piping, as shown in Figure 11-5.

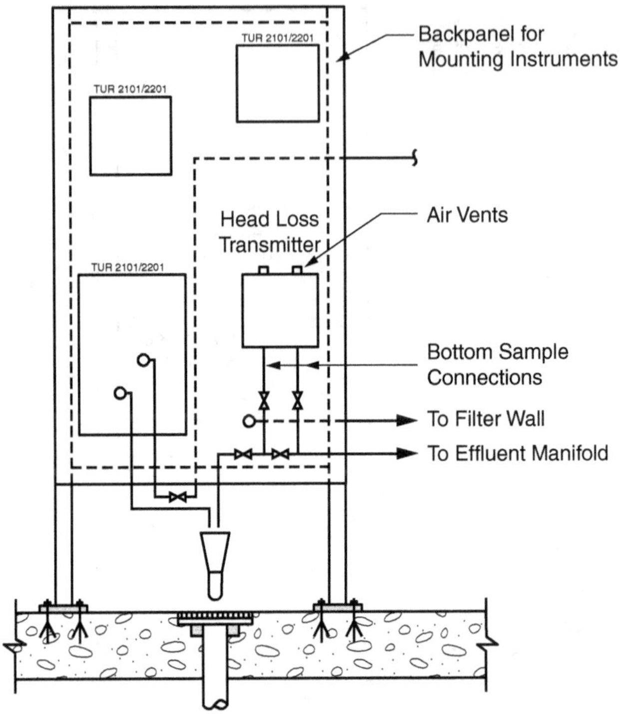

Figure 11-6 Head loss transmitter mounting detail

HEAD LOSS TRANSMITTER MOUNTING AND CONNECTIONS

Proper mounting and routing of connections are also important in obtaining accurate head loss information.

Mounting

Differential pressure transmitters are commonly used for water, as well as other uses such as for steam, air, and some chemicals. The transmitter body typically has two process water connections on one end (high and low pressure) and two air vents on the other. Depending on the application, the mounting requirements may be different. When used for water, the air vents should be on top and the sample water connections on the bottom, as shown in Figure 11-6. With this orientation, any air in the sample connections can be vented upward by opening the vent screws.

NOTE: The head loss transmitters have been mounted upside down prior to startup in many new installations. It is important to read all the instructions for these transmitters, including the fine print.

Sample Connections

To keep air out of the transmitter during normal operation, it is recommended that a low loop be constructed in the sample tubing, as discussed previously and shown on Figure 11-6. The sample tubing should be routed downward from the process piping, over to the transmitter, and then up into the body of the transmitter as shown.

SUMMARY

Accurate data are an obvious necessity for proper operation of any plant. It is, therefore, recommended that the operators make an extensive investigation of each separate piece of instrumentation and equipment to verify that the installation is correct.

CHAPTER TWELVE

Response Time

When the water quality or other data changes, it is important in plant operations to be able to detect these changes as soon as possible. Process response/detention time for instrumentation is defined as the elapsed time from when a change in the water quality occurs to the time it is detected by the instrumentation and recognized by the controls. There are several factors involved in making up the overall response time.

PROCESS RESPONSE/DETENTION TIME

In most treatment plants, chemicals for coagulation or other purposes are often fed into the process piping near the front or head end of the plant. The process response/detention time is defined as the elapsed residence time that the chemicals are in the process stream until changes can be detected in the instrumentation after treatment (filtration in this case).

When chemical coagulation is required, the necessary detention time is a function of the raw water quality, the water temperature, and the chemicals used. The detention time required should have been determined during the pilot study and then incorporated into the design of the plant. The actual detention time available in the plant should be the worst case required, which is often during cold water conditions and high process flow rates.

NOTE: The treatment plant may be difficult to operate and chemical usage may not be efficient, if adequate detention time is not available.

MEASURING PROCESS DETENTION TIME

The actual detention time available for coagulation may be determined by calculating the total volume of water between the point of chemical injection and the instrumentation. The volume will include process piping as well as the treatment units. However, the actual effective volume available may differ depending on plant flow rate and potential short circuiting.

It is recommended that the actual time be measured manually with a clock from the time the chemical feed rates or process rates are changed until the results are noted in the instrumentation.

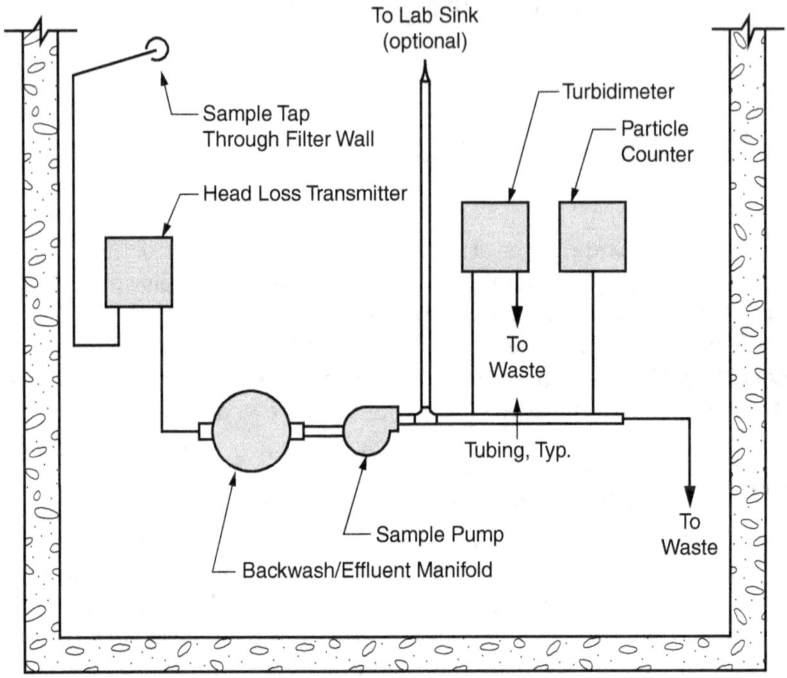

Figure 12-1 Typical instrumentation installation

It should be noted that the measured time may include a short period for the change in water quality to become uniform throughout the process piping in the filtration effluent. The instrumentation will not detect a change until water containing the different water quality passes by the instrumentation sample tap. Time delays of this type or short circuiting are more likely to exist in larger plants that have larger piping, especially at lower flow rates. However, the time for this to take place will also be part of the overall measured elapsed time. Any differences between the measured time and the calculated time may include this factor.

SAMPLE TUBING/PIPING DETENTION TIME

Once process water enters the sample tap, there is a time delay before the sample reaches the instrumentation. The time delay depends on the tubing diameter, the length of the sample piping or tubing, and the sample flow rate. A typical illustration of the installation of some types of instrumentation is shown in Figure 12-1. A typical calculation of sample detention time for these instruments is shown in the following section.

Assumptions

a. Assume 20 ft of ½-in. copper pipe/tubing.
b. Assume a sample flow of 500 mL/min for a turbidimeter.

Calculations

The detention time in the sample piping/tubing would then be approximately two minutes, plus a small amount of additional time to account for the volume in the sample pump(if used).

Recommendations

To minimize this time, the turbidimeter should be located as close as possible to the sample tap. Many regulatory authorities require the instrumentation to be close to the sample tap anyway, to detect changes in water quality as soon as possible.

INSTRUMENT INTERNAL DETENTION TIME

Depending on the type of instrument, there may be a detention time within the body of an instrument before changes are detected. For example, within the body of a typical turbidimeter as shown in Figure 12-2, there is a detention time of approximately 5 minutes at a sample flow of 500 mL/min. The detention time in other types of instruments may be considerably different. The manufacturer's literature should be consulted for more accurate information.

CONTROLS RESPONSE TIME

Depending on the software, programming, and to some extent the type of wiring system, the plant controls can also be a source of time delay. In a treatment plant, there are many devices reporting to, and controlled by, the plant control system. These devices may include instruments, process control valves, pumps, blowers, or other devices.

Sequential Polling

In older systems, the computer might *poll* or read the data from one device at a time. Then, after having reviewed all the devices, the computer system starts over again. In one such system, it took at least 2 minutes for the control system to acknowledge a command and then

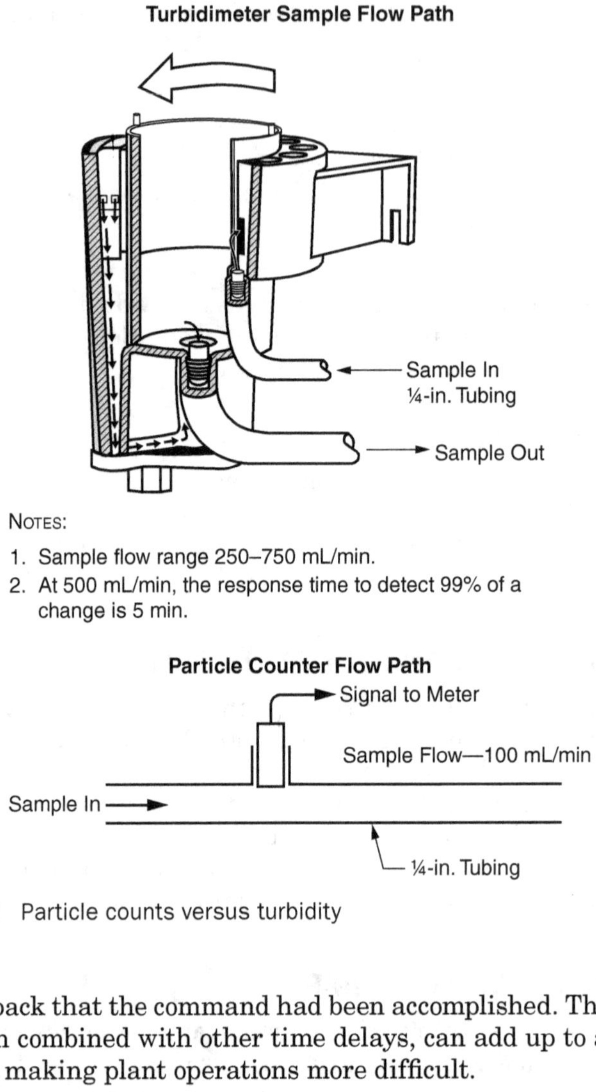

Figure 12-2 Particle counts versus turbidity

to report back that the command had been accomplished. These 2 minutes, when combined with other time delays, can add up to a substantial delay, making plant operations more difficult.

Report by Exception/Device Net

Using the Device Net communications system, all instruments and devices are essentially connected to a *party line* wherein any operational changes, such as a variable level or a valve opening or closing, can be reported or initiated nearly instantaneously as changes occur. Any device can report or activate at any time, while those that are inactive remain silent. A background *heartbeat* is used constantly to inform the plant controller that the devices are functional.

A Device Net system was illustrated previously in Figures 9-1 and 10-1. It is recommended that any new control system be at least equivalent to, or better than, a Device Net communications system in regard to response time. A control response time delay of less than two seconds (plus or minus) is recommended as a goal.

Example

At one treatment plant, the control system was able to print out all the pertinent data in the entire plant every 2 to 3 seconds. There are still instrument and tubing delays but practically no delays due to controls. Data of that quality were a great assistance in troubleshooting that plant.

Ethernet

An Ethernet communication system is similar to Device Net, as applied to larger systems/plants. In simplistic terms, Ethernet uses a series of switches, or modems, to route and collect the data before it is sent to the central controller or base station. The result is a much faster way of collecting and reading the data and a faster control system.

SUMMARY

When evaluating the effect of process changes, it is important to determine the various response times in order to determine exactly when each event occurred. For example, in the previous calculations, there is a response time delay of 7 minutes in the sample piping and the turbidimeter, plus a potential delay in reporting of 2 minutes. Even if the total process detention time is 60 minutes, a 9-minute delay would be a significant portion of it and can result in the production of that many minutes of poor quality water before a change can be detected. Therefore, it is in the operator's best interest to decrease the detention/response time as much as possible.

CHAPTER THIRTEEN

Operations Manual/Records

Keeping good historical records and having a good, useable operations manual are extremely important in being able to operate a treatment plant properly.

HISTORICAL RECORDS

A database of historical records should be maintained regarding the operation of the treatment plant. It is recommended that the historical data include the information contained in appendix B, as a minimum. Other information may be added as needed for the specific site.

Having this information available to the operator will help quickly identify the proper chemical feed settings when the raw water quality changes, by referring to data for similar water conditions that occurred previously.

Historical records should also include previous bench-scale/jar testing, as well as records of process changes that were made under different raw water conditions and during different times of the year. The records should include the following.

Raw Water Quality

A graph of each of the following data over time is recommended. Data from different seasons should also be included. It is recommended that all such data be filed according to the raw water pH and filed therein by date.

- Turbidity
- Color
- Alkalinity
- Water temperature
- pH

This data must be organized according to the needs of the particular plant. For example, turbidity may be more important than alkalinity at some plants or vice versa.

Process Description (for the Plant Operation on the Date in Question)

- Chemical settings for all pumps
- Plant flow rate
- Graphs of data from a filter run in an optimum condition
- Head loss
- Effluent turbidity
- Particle counts (if available)
- Typical clean bed head loss
- Typical filter run time
- Seasonal SCM settings (Please note that SCM settings may change, at least seasonally, when there are major changes in the raw water.)

NOTE: Refer to appendix B for a typical data sheet. Information on operational history and records is also contained in Reference 1.

Bench-Scale/Jar Testing

Even with historical records, it is recommended that bench-scale/jar tests be performed, before making any changes, to verify the recommended settings under the existing conditions. Records from these tests should also be included in the database.

Examples

At one plant, the traditional filter aid polymer had to be changed quickly from a cationic type to a nonionic type when the water became colder and a significant change in the alkalinity occurred. The need to change polymers occurred suddenly and was unexpected. In addition, there was no way to measure alkalinity at this plant. The change occurred because of a flood in the watershed that exposed different soil types. Once the operators recognized this condition, they were ready when it occurred in the next year.

At one small plant, the operators historical records consisted of a series of "post it" notes on a window. It is doubtful if this type of record keeping was beneficial to anyone.

Operator Training

Another benefit in maintaining historical records is in operator training. One very complex plant had the capability of feeding six or eight different chemicals in each of six or eight different locations. When asked how the plant was operated, the chief operator said that he had been working in that plant for over 30 years and knew exactly what to change at any given time or with any given set of raw water conditions. The historical records were there, but they were all in the operator's head. However, it is important to have written records for use by others. A new operator would be lost in such a complex plant. Having historical records and an operational guide would be highly beneficial for new personnel to review.

Operator Turnover

In a large plant, such as previously discussed, there would be numerous operational personnel. It is hoped that some of them would be trained to take over the supervisor's job in an emergency or if the chief operator retired or became sick. However, when a new operator, or new chief operator, has been brought in from the outside, it would be very helpful to have historical records for reference. It would greatly speed any transitional process. New operators are very common, especially in smaller plants.

Time Required

Obviously, the development of historical records takes time out of an operator's day, but in the long run, it may be cheaper to do it than to not do it. There have been plants that almost depleted the system storage of water while trying to cope with changing conditions. Data should be recorded every day, and at least a year's data should be accumulated to cover all of the seasons.

As with the recommended maintenance procedures, recording this data requires the operators' time, and smaller plants may not have the necessary personnel. Many treatment plants, small ones included, are being built with computerized controls and SCADA systems (supervisory control and data acquisition). If this is the case, most of the data can be generated by the control system, with minimal input required by the operator. The control software would then have to be written to organize and store the data in a useful manner.

Troubleshooting

Historical data will also help in the detection of potential problems. For example, if the clean bed head loss is increasing over time, the filter media or membranes (if used) may be fouled. An investigation should then be made to determine the cause and any remedial action that should be taken.

Refer to Reference 1 for troubleshooting assistance.

NOTES: Everything should be recorded—chemical feed rates, raw water condition, and treatment efficiency, as well as any events that occur (rainstorms, equipment failure, etc.). It will be of great benefit in the future for the reasons previously discussed. The process of preparing the data will also add greatly to the operators' own knowledge.

OPERATIONAL MANUAL/GUIDE

Once the plant design is complete, the most important task for the designer is then to write a *useable* operational guide for the owner/operator. The guide should be available before the plant is operational. Otherwise, it may be difficult for operators to learn how to run a new plant.

Typical discussion items for an operational manual are:

- Does all the plant equipment start/stop all at once automatically? If not, what is the necessary procedure?
- Do various components have to be started/stopped separately?
- Are the various supply pumps, chemical pumps, and process pumps started/stopped automatically? Are valves closed/opened automatically?
- What is the exact sequence for starting and stopping the plant or the backwash cycle? For example, does one valve open completely before the next valve actuates? Knowing the exact sequence of events can be extremely important in determining their effect on performance data and in troubleshooting.

Example

For example, at one plant several of the chemicals began to siphon into the system when the plant was shut down. The isolation valves had to be shut off individually until the problem was resolved. Sample lines drained a filter at another plant when the filter was left off for a period of time. Then, when the plant was restarted, those valves had to be reopened.

Valve Fail Position

- What is the fail position for the valves? Is it open or closed or right where they are at the moment of failure?
- What happens during an electrical failure?
- Are they electric, hydraulic (water), or pneumatic (air) actuated valves?

Example

One large plant overseas had all pneumatically actuated valves. None of the valves worked because someone had stolen the air compressor and sold it on the black market. It is amazing that they could operate the plant at all. Obviously, this is an extreme case, but the principle still applies.

- During startup, do the plant components (filters for example) start incrementally, or all at once? NOTE: Startup sequences vary widely. Larger plants may start incrementally, while the filters in smaller plants may start all at once.
- How is plant flow controlled during the startup transition?
- What controls the transition from partial to full flow? How long does it take? How accurate is the chemical feed pump control during this process?
- Does the plant flow rate vary, or does it normally operate at fixed speed?
- Are the chemical pumps flow paced or are multiple pumps used?
- Do the chemical systems have positive shut-off control? At one plant, the chemical systems backsiphoned into the supply well when the plant was shut down.

Type of Control System

The type of communication and control system in use at a particular plant can have an impact on how well the plant operates. Some systems used include the following:

- **Direct Wire (old style)**—all components are wired to the plant controller individually.
- **Distributed Control (common)**—common data wiring uses either Device Net or Ethernet.
- **Radio or Telephone System Control**—the type of communication system used depends on the area covered by the water system and previous history. For example, some systems have frequent telephone problems and use more dependable radio communications, which the operators control.

Type of Filter Control

The startup and backwash sequencing is different for the various, different types of filter control discussed previously. The primary filter control types are listed below.

- Equal loading
- Declining rate

- Equal rate
- Equal loading/variable level

The operators need to be very familiar with the type of filter control. The sequencing is different and the valve operation is different. The operators need to know the difference to be able to better operate the plant and to conduct troubleshooting tasks.

Example

At another plant, no one knew exactly how the system operated. In that case, the operators relied completely on the automatic controls. If those controls were ever to fail, which could happen, the operators would be at a complete loss in knowing how to operate the plant. Refer to Reference 1 for an extended discussion of filter control types.

OBTAINING A USEFUL OPERATIONAL MANUAL/GUIDE

Obtaining a good useful operational guide is rare. Many engineers typically use the entire budget in their contract to do the design, and then leave the owners to their own devices or with only an equipment maintenance manual. In fact, there are plants where the designer could not even operate the plant. How then are the operators to learn? It is, therefore, up to the owner to insist that an adequate budget be allowed to obtain a usable, site-specific operational guide, and that an adequate description of their needs in this regard be included in the contract language. Otherwise, it can be up to the designers to interpret what is to be provided.

The owners should know what they want, and then take the necessary steps to obtain a quality product. Visiting as many other plants as they can to see what other plants do and have can be a substantial benefit.

SUMMARY

This section describes the information that should be available to allow plant operators to better understand their plant. This section also discusses the type of information that should be available in case the operators need to write their own operational manual, which may be the best in the long run. Even then, some outside assistance may be necessary.

CHAPTER FOURTEEN

Evaluation of Process Data Over Time

Interpreting the cause and effect of changes in the operating parameters of a treatment plant is a necessary function for operators. They need this skill to troubleshoot process problems that occur, as well as in making normal operational adjustments.

The previous chapters of this text have discussed various aspects of monitoring data and have provided background material for that purpose. This chapter discusses the evaluation of process data.

NOTE: The emphasis of the following discussion will be on changes or trends in the data over time.

TURBIDITY VERSUS TIME

First, assume a graph of turbidity versus time, which might be typical of the effluent of a granular media filter. The solid line on Figure 14-1, illustrates a filter-to-waste cycle that might occur after a backwash, followed by a gradual rise in turbidity until it again reaches a terminal value, at which time another backwash is initiated and the cycle starts over.

No values are shown on this graph as the only interest is in changes or trends over time.

Filter-to-Waste Cycle

After a backwash cycle or when a filter is brought on line from standby mode, a filter-to-waste cycle is initiated. It is similar to the filtration mode except that the effluent or product water is diverted to waste until the filter is conditioned properly and is making high quality water. In a typical filter-to-waste cycle, the turbidity ramps up quickly and then is reduced to an acceptable value as the media settles and the filtration efficiency improves, at which time the filter production run begins.

In evaluating a filter-to-waste cycle, the important values are the maximum turbidity level reached and the length of time it takes for the turbidity to be reduced to an acceptable value for production. Any significant variation from these values over time may indicate an

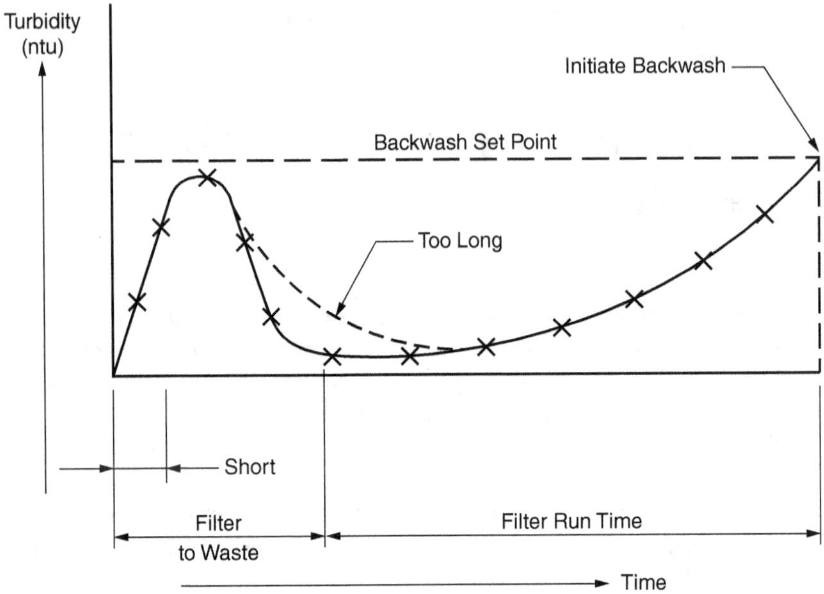

Figure 14-1 Turbidity versus time typical of a granular media filter

operational problem and should be evaluated and any appropriate process changes made accordingly.

Filter Run Time

The second portion of the graph in Figure 14-1 represents the filter run time. It should begin at a similar turbidity level for each filter run, and the actual operational time before reaching the terminal turbidity set point should be similar, unless there is a change in raw water quality.

CHANGES IN THE FILTER-TO-WASTE CYCLE

Changes in the filter-to-waste cycle may include too long a time period required to achieve the proper water quality or too short, each of which has different causes. Changes of this type are illustrated on Figures 14-1 and 14-2.

Long Filter-to-Waste Cycles

A longer than normal filter-to-waste cycle (Figure 14-1) may be due to weak floc, upset media or support gravel (if used), a higher process flow rate, or the filter media not being properly cleaned.

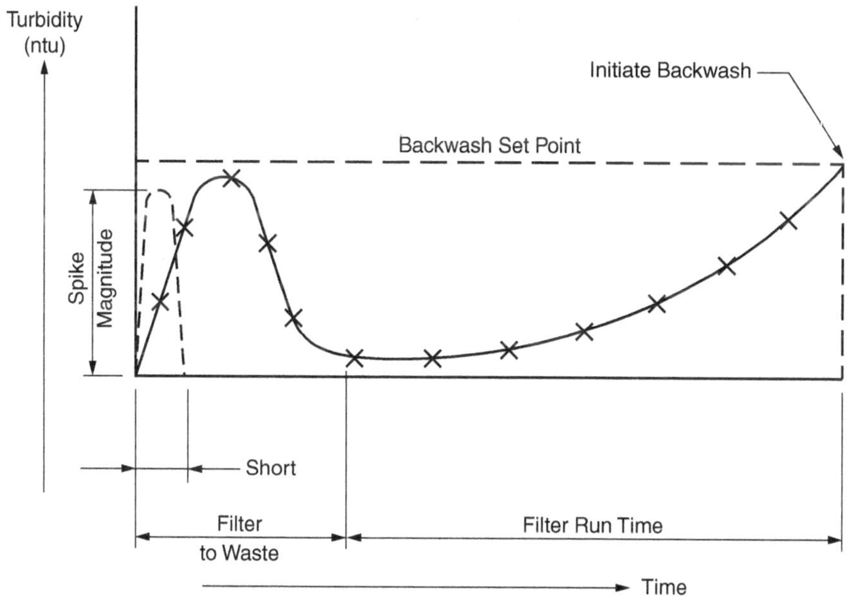

Figure 14-2 Changes in the filter-to-waste cycle

Weak floc. Weak floc can occur as the result of changing raw water quality, assuming that the floc strength was good previously. Weak floc may not be able to withstand the shear that occurs in the media bed because of the acceleration of water around the media particles. Therefore, a longer filter-to-waste cycle may be required to condition the media properly. Refer to Reference 1 for an extended discussion of this issue.

Floc shear. Granular media typically occupy approximately half of the volume in the space in the filter. Therefore, when the downward flow of water penetrates the surface of the media, the water velocity has to accelerate by a factor of 2 to maintain the process flow rate. The resistance of the media to this acceleration is also responsible for the clean bed head loss, which tends to tear up or shear the floc into smaller particles, which may not be easily filtered out. Toughening the floc to be able to withstand this shear is often accomplished by using a filter aid polymer.

Additional head loss (operational head loss) is developed over time, as the filter accumulates solids. Refer also to Reference 1 for an additional discussion of filtration efficiency.

Changes in floc strength. Changes in floc strength, assuming it was good initially, can be caused by a number of issues such as rain storms or other seasonal changes in the raw water quality. In such cases, the changes in raw water quality should be monitored by the plant instrumentation and may take the form of changes in any of the following:

- Turbidity
- Suspended solids
- Color
- Temperature
- pH
- Alkalinity
- Silt density index (SDI)

Process changes. Changes in these values can take place rapidly in the raw water. When any changes of this type are detected, the plant water chemistry may also need to be changed including the coagulant, polymer(s), and buffering chemicals. If the plant instrumentation includes a streaming current monitor (SCM), it may provide the operators with the proper information to make the necessary process changes. Whatever the cause, it is recommended that changes in the process be verified by bench-scale/jar testing.

On-line instrumentation. On-line instrumentation is a great benefit to the operators and is a case where more may be better than less. The operators can use all the information that can be made available. However, too much emphasis is sometimes placed on having automatic control of the treatment process using on-line instrumentation, all of which is subject to failure or which may also respond incorrectly to unusual conditions. The operators should use on-line instrumentation as a tool so that they can make the best process decisions. If automatic process control is used, it is recommended that the operators closely monitor any changes that take place.

Quality control—example. One plant that was not manned 24 hours a day, used an SCM to automatically control the coagulant feed rates. An unusual alkalinity condition occurred that essentially "fooled" the SCM into maximizing the coagulant feed and resulted in filling the treatment plant full of coagulant. Therefore, it is recommended that jar testing always be practiced to verify the results of the automatic instrumentation, especially because the public health is involved. There is no substitute for quality control.

Upset media. If changes in the values previously listed occur slowly over time, it may be possible that issues other than raw water quality may be involved. For example, if granular filter media are not being properly cleaned, mud-balls may develop over time resulting in loss of treatment efficiency. Periodic evaluation of the filter media is recommended to provide an early warning of any such problems.

Refer to Reference 1 for recommendations regarding filter media maintenance.

Changes in process flow rate. Treatment plants may operate in many different modes. Where variable process flow conditions occur, the treatment efficiency may change with the rate. Lower flow rates typically result in higher filtration efficiency, higher quality product water, and longer filter runs. Conversely, it is possible that increasing the flow rate can result in less efficient filtration efficiency, poorer water quality, and shorter filter runs.

Proportional flow control. If a plant is designed for proportional flow control (chemical feed rates being modulated automatically in proportion to flow), it is necessary for the chemical pumps to respond accurately to changes.

Linear or nonlinear. Proportional flow control of chemical pumps may be linear. That is, if the flow changes 10 percent, the chemical feed pumps also are changed by 10 percent. However, chemical pumps may not always actually operate in a linear manner, especially if they are slightly fouled or are pumping a viscous chemical. Therefore, it is recommended that all chemical feed pumps be maintained and cleaned frequently. In addition, periodic flow calibration of each chemical pump is recommended to assess the accuracy of the control system.

In some cases, it may be necessary to make slight adjustments to the chemical feed rates after the controls have made the automatic adjustments according to the flow rate change. For example, if alum is used as the coagulant, there will be an optimum pH value for coagulation. If the flow control does not exactly hit the desired value, manual changes can be made. An SCM can also be used for this purpose.

Long filter-to-waste summary. Filter-to-waste cycles that are longer than normal are mostly indicative of one or more problems. When a filter has normally operated in a similar manner for a length of time, any significant changes in the time required for the filter-to-waste cycle should be investigated immediately and rectified if necessary, even if these changes were slow to occur. A quick response is especially important because longer filter-to-waste times may mean a degradation of the filter media. For this and other reasons, it is strongly recommended that historical records be maintained to be able to compare current operational conditions with those when the filter may have been newer and properly optimized.

Shorter Filter-to-Waste Cycles

Shorter filter-to-waste cycles can occur if there is an accumulation of fines on the surface of the media, excessive polymer feed, heavy floc, and possibly low unit flow rates. Several of these issues will also result in shorter filter runs.

Accumulation of fines and skimming. Another media problem that can occur over time is the accumulation of fines as anthracite (if used) breaks down. Fines on the surface of the media will cause *surface filtration* with rapid head loss buildup and will prevent the solids from penetrating into the media.

A buildup of fines may be characterized by both shorter filter-to-waste cycles and shorter filter runs. A simple investigation of the media surface can determine if there are fines that need skimming. Again, refer to Reference 1 for skimming recommendations and procedures.

Excessive polymer feed. An excessive polymer feed can easily blind off a filter and result in both short filter-to-waste cycles and short filter runs. Excessive polymer feed rates can sometimes be verified by a visual examination of the filter media. However, by that time the damage will be done and the media may be ruined. It is better to determine the proper feed rates by bench-scale/jar testing and by frequent calibration of the chemical pumps.

Example. The water over the filter in one plant during startup was so heavy with polymer that it felt like syrup. The result was that the media were fouled and had to be replaced. More is not always better. It should be noted that the equipment manufacturer was responsible for this condition during startup, because of its rush to produce good water. Plant operators need to be very knowledgeable.

Strong or heavy floc. Strong or very heavy floc can blind off a filter rapidly and possibly result in a shorter filter-to-waste cycle. Floc of this type could be caused by excessive polymer (discussed previously) or inefficiency of or overloading of the clarifier/sedimentation basin (if used). Refer to Reference 1 and the following sections in this text for a discussion of clarifier/sedimentation basins.

FILTER RUN TIME

Any changes in the normal filter run time are of great importance to the operators. As a general rule, a filter run time of at least 24 hours should be the goal. Shorter time periods will have the result of increasing the percentage of backwash waste and reducing the overall volume of product water.

A typical graph of a shorter filter run due to high turbidity is shown on Figure 14-3. The figure shows the filter-to-waste cycle to be normal. However, if the filter run is shorter than normal, it is also likely that the filter-to-waste cycle will be affected as discussed previously.

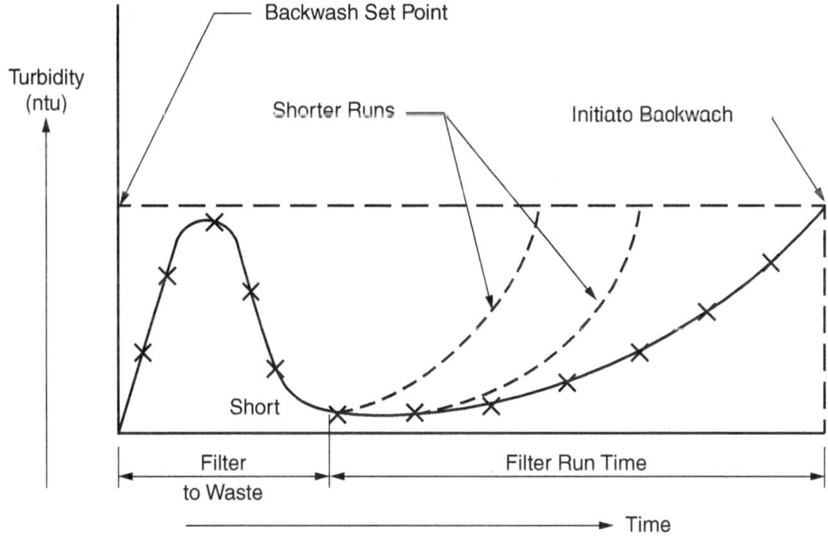

Figure 14-3 Turbidity versus short filter run

Shorter Filter Runs

Short filter runs can be caused by of a number of different issues including the following:

Changes in the raw water quality. If changes in the raw water quality (either for better or worse) are not taken into account, the treatment efficiency can deteriorate. In those cases where the effluent quality is reduced, it is recommended that the operators first check for changes in the raw water quality. Then, bench-scale/jar testing should be done, and/or SCM or other instrumentation should be used to determine the cause and to help in determining changes that need to be made.

Higher filter flow rates. When the process flow is increased, the unit filter flow rate will increase accordingly, resulting in greater head loss and floc shear with the potential for driving particles through the filter and also causing a reduction in effluent quality. When that occurs, the head loss will build up more rapidly resulting in shorter filter runs. There may also be a period of time where the water quality is worse. If it recovers quickly, the filter run can continue. However, if the water quality stays bad, the filter run may have to be terminated.

NOTE: There will be a maximum effective unit flow rate for the filters above which the water quality and run time cannot be maintained.

The initial pilot study should determine what the maximum rate is, and the plant should be designed accordingly.

Clarifier/sedimentation basin inefficiencies. As mentioned previously, inefficiency in solids removal in the clarifier/sedimentation basin can result in heavier than desired solids being sent to the filter. These inefficiencies can result from excessive solids buildup and/or design limitations. In either case, solids can usually be observed in the clarifier/sedimentation basin effluent if there is a problem. Refer to Reference 1 for a more extended discussion of these issues.

There have been numerous treatment plants with design limitations in the pretreatment or clarification/sedimentation basins. If this is the case, the operators need to determine the optimum operating conditions for their plant equipment. Operation in excess of the optimum conditions will probably result in shorter filter runs. Another option is to determine if mechanical/physical improvements can be made to improve efficiency.

Upset or fouled media. Upset or fouled support gravel and/or granular media are illustrated in Figure 14-4, which shows disturbances in the normal gravel layering. An upset condition can also include ruptured porous plate caps (where there is no support gravel), as well as broken or ruptured membranes.

Upsets nearly always occur in the backwash cycle and are often caused by some type of control problem. Membranes can rupture as a result of fouling or becoming brittle over time.

Refer to Reference 1 for additional information on upset conditions and troubleshooting.

Low backwash rates.

- The proper backwash rate for granular filter media depends on the water temperature and the effective size (ES) and uniformity coefficient (UC) of the media (that of the anthracite being the controlling factor). The actual backwash rate required should be determined by pilot testing and/or expansion tests at the end of the backwash cycle. Refer to Reference 1 for a more complete discussion of backwash rates and measurement.
- Many plants use the same rate much of the time. In that case, the rate may be too low during the warm summer months and too high during the winter.
- If the rate is too low, the media may not be cleaned properly and may become fouled. Fouled media lose filtration efficiency and often reach terminal turbidity in the product water sooner than a clean filter, resulting in shorter filter runs.

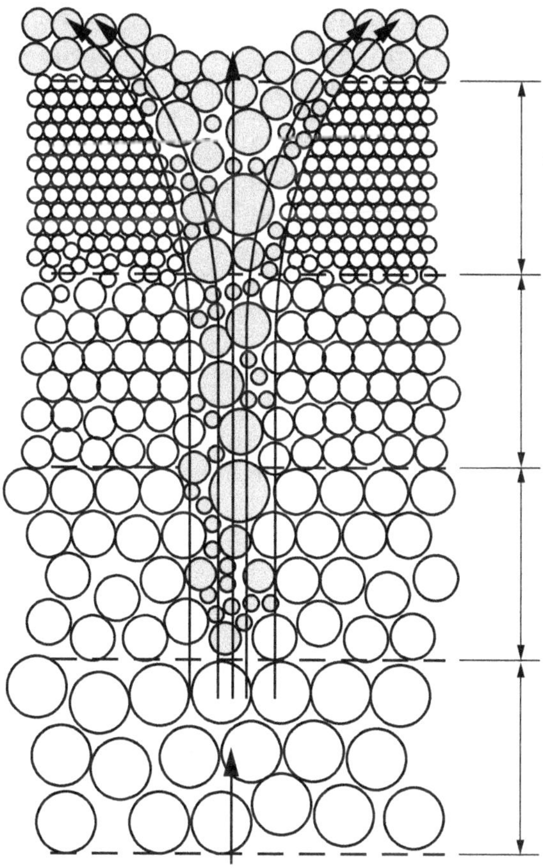

Figure 14-4 "Blown" single taper support gravel

High backwash rates. Using higher backwash rates does not necessarily mean that the media will be cleaned better. When backwash rates are used that are higher than necessary, it often results in some of the media washing out of the filter and being lost. Excessive media loss will likely result in less solids storage, poorer treatment efficiency, and shorter filter runs.

Loss of media can be detected in the backwash wastewater at the end of the cycle when the water becomes clear. Media grains can then be seen washing out if that problem is occurring. Loss of media is a common problem in many installations. Refer also to Reference 1 for additional information and a troubleshooting guide for this issue.

Entrained air. Some surface water sources contain high amounts of dissolved air. Cascading raw water in the delivery pipeline or free fall in the treatment plant can also cause this condition. Then, when the air saturated water passes through the filter, the associated head loss reduces the pressure and causes this air to come out of solution. Small bubbles can then attach to the media grains and cause an artificial buildup of head loss, resulting in short filter runs. These bubbles will cause some of the media to float out of the filter during backwash. Here again, loss of media can also result in treatment inefficiencies.

Entrained air will look like small air bubbles covering the water surface during backwash. Once a plant is constructed, the only practical solution to eliminate entrained air is to shorten the backwash cycle to eliminate negative pressure conditions within the media during filtration. If this condition is expected to exist, the filter should be designed initially to be deeper to mitigate the problem. A minimum height of 6 ft of water over the media, during filtration mode, is recommended for this purpose. Refer to Reference 1 for an extended discussion of this subject.

Long Filter Runs

Filter runs longer than 24 hours generally mean that the treatment plant is working well, or that the raw water quality is high, or both. It also means that the percentage of wastewater loss is less, as is energy usage. Other factors include lower filter rates and high clarifier/sedimentation basin solids removal efficiency. All of these are good and indicate very efficient operation. However, it is recommended that the operators be very careful if the runs are significantly longer than 48 hours. Very long runs can result in compaction of the media and potential fouling.

HEAD LOSS VERSUS TIME

The buildup of head loss over time is an important factor in any filter. A typical graph is shown in Figure 14-5. The graph starts when a filter is clean and builds to a point (terminal head loss) where the filter run is terminated and a backwash procedure is initiated.

Several factors regarding head loss are discussed the following sections.

Total driving head. The total driving head is the pressure or height of water that is available to operate the filter. The total driving head available must be at least equal to the sum of the clean bed head loss and the design operational head loss.

Figure 14-6 also illustrates the relationships between positive and negative driving head, the sum of which is the total driving head.

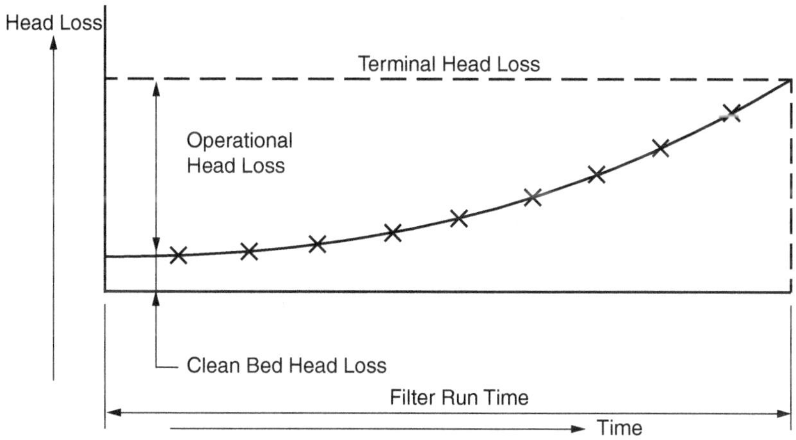

Figure 14-5 Head loss versus time

Table 14-1 Typical clean bed head loss

Filter Component	Approximate Clean Bed Head Loss During Filtration
18 in. of anthracite	9 in. to 12 in.
12 in. of filter sand	12 in.
Support gravel	Minimal
Filter underdrain	Minimal
Filter rate control valve	12 in.
Flowmeter	6 in.
Piping	6 in.
Total	Assume 4 ft of clean bed head loss

Refer to Reference 1 for a more extensive discussion of positive versus negative driving head.

NOTE: A high amount of positive driving head is desirable if there is a potential for entrained air.

Clean bed head loss. Table 14-1 contains a typical calculation of clean bed head loss. In Figure 14-6, the clean bed head loss represents the starting value at the beginning of a filter run and at the operational flow rate for granular media.

NOTE: The clean bed head loss varies according to the flow. For example, as the operational flow increases the clean bed head loss increases accordingly.

Operational head loss. The operational head loss represents the pressure available (or height of water), over and above the clean bed head loss, for the accumulation of solids while maintaining the operational flow rate. It is recommended that the operational head loss component not be greater than 10 to 12 ft of water. Pressure filters may have greater pressure available, but it is recommended that they be limited to the same value also. Higher operational head loss can lead to compaction and fouling of the media.

Example. One plant had pressure filters where the operational head loss was very high, and there were 3-ft diameter mud-balls in the media.

Gravity Filter

In a gravity filter, the total driving head is the vertical distance in feet from the operating water level over the filter down to the clearwell level, assuming that the discharge into the clearwell is submerged or sealed. Refer to Figure 14-6 for an illustration.

Pressure Filter

The available pressure minus the necessary discharge pressure determines the total driving head available for a pressure filter. However, it is still recommended that the operational head loss be limited to 10 to 12 ft of water. Even though the filter is a pressure filter with greater head loss available, the physical aspects of solids removal and media operation do not change.

Operational Mode

Depending on the operational mode of the filter (constant rate, variable level, etc.) and the design of the clearwell (constant level or variable level), the total driving head available may change during the filter run. If that occurs, it is possible that the head loss versus time curve may not be as smooth as shown in Figure 14-6, and there could be some irregularities or blips in the data. If there are such blips, the operators need to determine whether there are any modifications or corrections needed to eliminate them.

VARIABLE DRIVING HEAD

In Figure 14-7, a filter system is illustrated wherein the clearwell level varies, which in turn increases or decreases the driving head. A variable level condition can occur in the clearwell, if there is no separate well for the final effluent pumps. As the amount of driving head

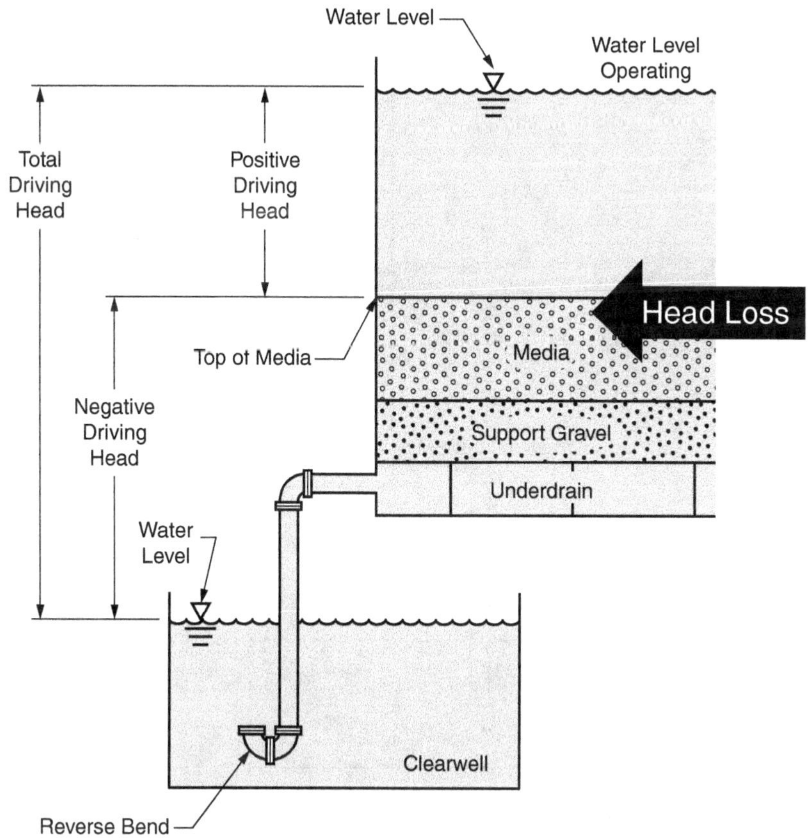

Notes:

- Positive versus negative driving head.
- Low positive—higher potential for entrained air.
- High positive—requires deeper filter.
- Most head loss occurs near the top of the media bed.
- Operational head loss:

	Total driving head
−	Clean bed head loss
	Operational head loss

- Gravity discharge is shown—pumped discharge has similar issues.
- Submerged discharge with reverse bend seal required.

Figure 14-6 Driving head

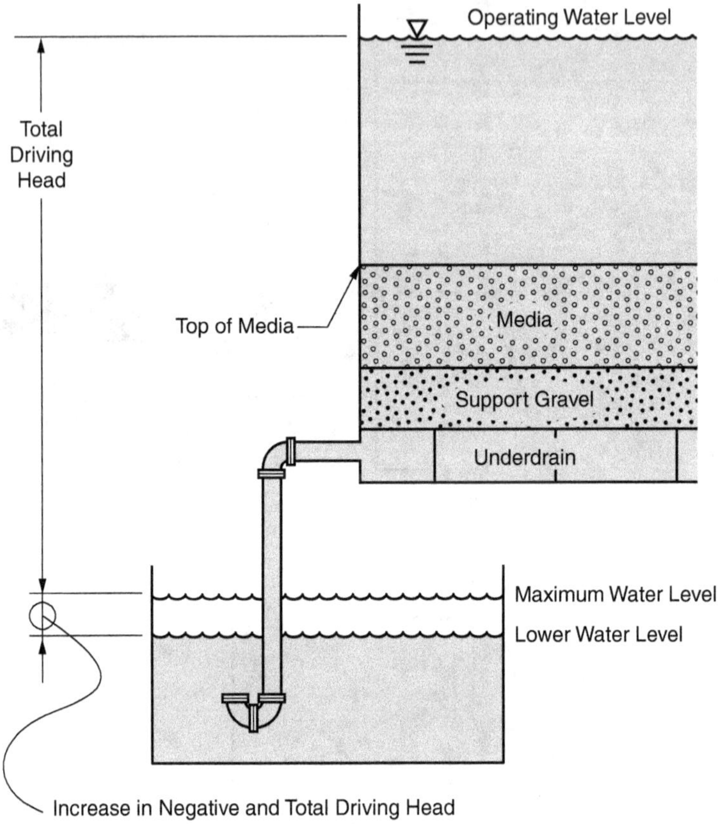

Notes:

- Clearwell level varies.
- Increase/decrease in negative and total driving head.
- Effluent flow varies.
- Constant control adjustments.

Figure 14-7 Variable driving head

changes, the flow rate may change also. Depending on the design of the system and the operational mode used, an effluent rate control valve may be used to maintain a constant operational flow rate. Nearly constant modulation of the valve would be required to maintain a constant rate. Depending on the control's response time and valve accuracy, driving head changes could cause irregularities in the head loss versus time curves and in the overall water quality.

Example. One treatment plant had a clearwell where the operational level varied widely. In addition, the effluent rate control valves

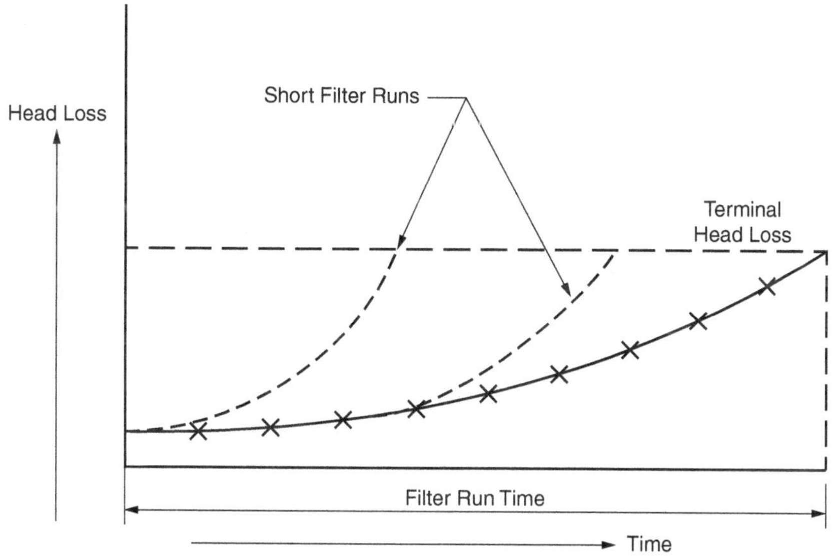

Figure 14-8 Changes in head loss versus time

had difficulty making rapid adjustments accurately, which in turn caused a wide fluctuation in head loss. Problems of this nature also caused considerable problems in maintaining the desired water quality.

Changes in Head Loss Versus Time

Changes in head loss versus time refer to conditions in which the terminal head loss is reached more quickly, or longer than, the normal conditions.

Shorter times. Shorter filter run times (Figure 14-8) caused by high head loss buildup can occur because of a polymer dosage that was too high, a buildup of fines on the media surface needing skimming, high solids overflow from the sedimentation basin/clarifier, a higher process flow rate, entrained air, or other issues.

An important part of plant operations is to be able to recognize when conditions have changed or vary from the normal. For this purpose, historical records are very valuable. Then, once this condition has been recognized, these issues can be evaluated for potential causes.

NOTE: High or rapid head loss buildup will also be characterized by shorter filter run times.

Longer times for head loss buildup. A longer than normal filter run caused by low or slow head loss buildup is usually due to improved raw water quality, improved clarifier/sedimentation basin efficiency, or

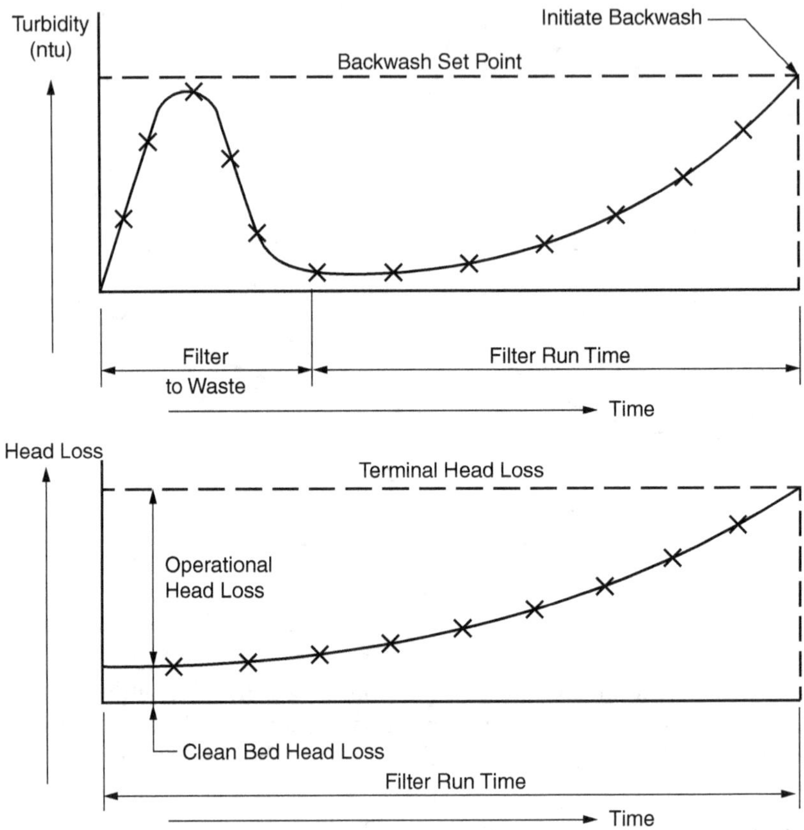

NOTE:
The optimum filter run would terminate on high head loss, just before terminal turbidity is reached.

Figure 14-9 Turbidity versus head loss

lower process flow rates. All of these are good, and the operators should continually strive to improve them. However, care should be taken that the filter run times are not excessive.

HEAD LOSS VERSUS TURBIDITY

An extremely valuable operational tool is the ability to either overlay or have both turbidity and head loss versus time curves displayed on the same computer screen (Figure 14-9). The ability to overlay other data is also valuable.

Changes or Irregularities

Having both sets of data on one screen provides the operator the following tools:

- Changes in head loss and turbidity can be monitored separately or together.
- Any changes with either one that occur over time can be monitored.
- Perhaps most importantly, the operator may be able to determine if any changes or irregularities in one affect the other. Examples of such changes/irregularities include:
  - — Higher than normal turbidity with little or no increase in head loss may mean that the filter media are upset.
  - — Higher head loss with little or no increase in turbidity may mean that the filter aid polymer dosage is too high, or the media may need to be skimmed.

In either case, it would be a great help to the operator to determine the cause of unusual conditions by a proper evaluation of the data.

Optimum Filter Run

The optimum filter run would terminate on the high head loss set point, just before the terminal turbidity set point is reached. It is generally not good if the terminal turbidity set point is reached before the high head loss set point. If that occurs, a higher filter aid polymer dosage may be required, or the media may need to be investigated for fouling or upset conditions.

Optimization

Optimization consists of several components including

- Having the right chemical dosages for the current raw water conditions
- Operating at the proper process flow rate for the current conditions
- Having the sedimentation/clarification basin operating properly
- Having the proper filter media design
- Having the media clean and in good condition

NOTE: Optimization is a continuing process, not any specific achievement or goal. In other words, the operational staff should continually strive for optimization of each piece of equipment and treatment process.

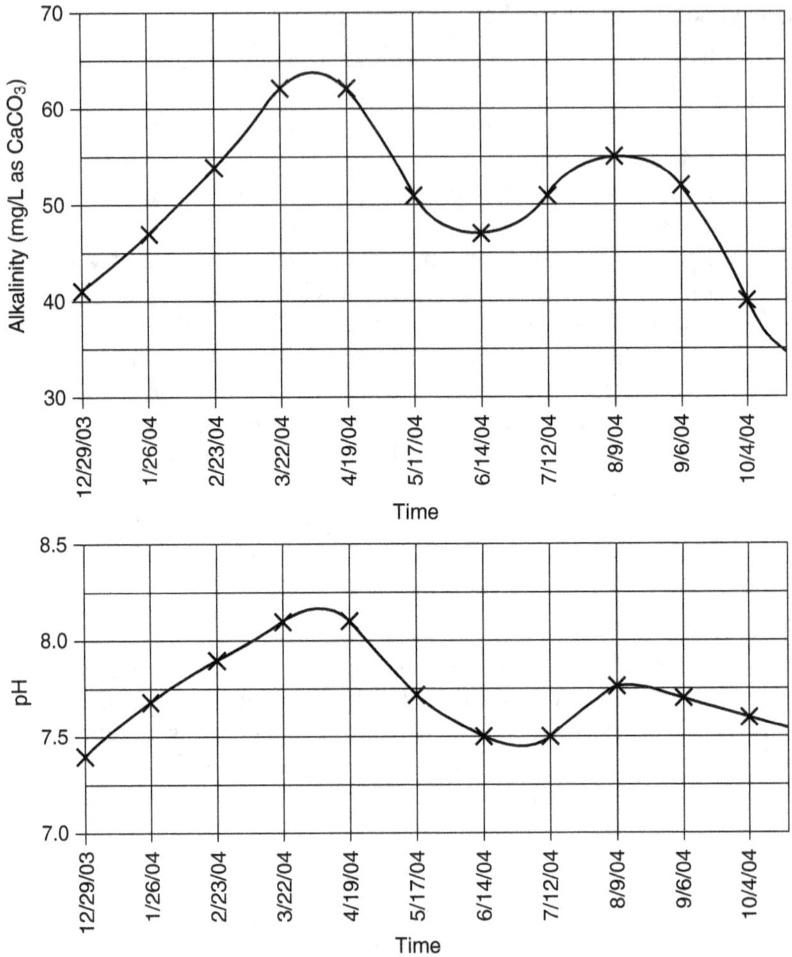

Figure 14-10 Alkalinity and pH versus time

ALKALINITY/pH VERSUS TIME

Both alkalinity and pH are shown on the same computer screen in Figure 14-10. It should be noted that the data on this figure are from a real treatment plant.

Variations

Variations in either type of data can be seasonal or can occur daily.

Seasonal variations in alkalinity and pH are common with surface sources. However, well water sources are more constant unless the groundwater level is drawn down and the well starts drawing water from further away.

Because seasonal variations may occur slowly, it is easier for the operators to anticipate and to make the process changes needed. Here again, having historical data on similar changes that occurred in previous years would be a great benefit.

Daily changes in alkalinity and pH are also relatively common in surface sources, especially those with periodic algae problems. When daily changes occur, making the appropriate process changes can be difficult. Several alternatives are noted in the following sections.

An automatic pH control loop should be developed using the coagulant or a separate buffering chemical or both. Implementing this option may require a coordination of both the coagulant and buffering chemical (if used). The use of an SCM is also recommended. An automatic response may be the best option in terms of best water quality and efficient chemical use.

An average chemical feed dosage could be used. Using this option may result in the production of poor quality water on both ends of the average.

If the raw water conditions allow, a worst-case chemical feed could be used all day. Using this option only works if reasonable water quality can be obtained on both ends of the average. This approach has been used in a number of plants, although it may be questionable in some cases.

Another option may be for the operators to make chemical feed adjustments manually every hour. Highly qualified operators would then be required on a 24-hour basis, which may be the best option in most cases, even if hourly process changes are not required.

pH Related to Alkalinity

Variations in pH often correspond with similar variations in alkalinity. On Figure 14-10 the trend in both is similar. On-line monitors for both are recommended.

Changes in pH may require a change in the coagulant dosage.

Changes in alkalinity may require a change in the buffering chemical used, as well as a change in the coagulant.

CHANGES AFFECTING FILTRATION EFFICIENCY

Figure 14-11 illustrates another set of data from a real operating treatment plant that compares alkalinity and turbidity versus time.

Rising Alkalinity

The alkalinity is rising at the beginning of the graph that compares to several rises and falls in the turbidity data. These changes represent significant seasonal changes that will require significant changes in the water chemistry. Knowing this trend from historical data will allow the operators to be prepared.

NOTE: These changes may also require changes in the operational set point of an SCM (if used).

Falling Alkalinity

The largest correlation in the data occurs when the alkalinity begins to fall, which relates to a turbidity spike. A comparison of weather data may indicate the beginning of a rainy season during this time. Please note that according to the timescale on Figure 14-11, the data are seasonal and therefore any response time in the treatment plant is not relevant.

Data Analysis

According to the data in Figure 14-11, the rise in alkalinity was apparently unexpected and resulted in the production of poor quality water until the situation was recognized and addressed. The resulting spike may have been short but should have been detected much sooner. Perhaps there was not adequate instrumentation at this plant.

Typical Evaluation

The analysis of data in these graphs is indicative of the type of monitoring and evaluation that can and should occur in a treatment plant when changes occur. It is hoped that operators at plants with similar water conditions will have the necessary instrumentation to monitor these changes and/or will have the historical data to help them anticipate them.

NOTE: The analysis of this type of variation in the raw water requires frequent alkalinity testing. Many treatment plants, especially smaller ones, may not have the capability of analyzing alkalinity, although turbidity data are more commonly recorded. However, the data demonstrate that on-line alkalinity instrumentation would be valuable to have, while manual testing is recommended as a minimum.

Particle Counts Versus Turbidity

Particle count data are extremely valuable and would be a great benefit to all treatment plants. The ability to compare particle count and turbidity data also can provide additional insights as to the function of a treatment plant.

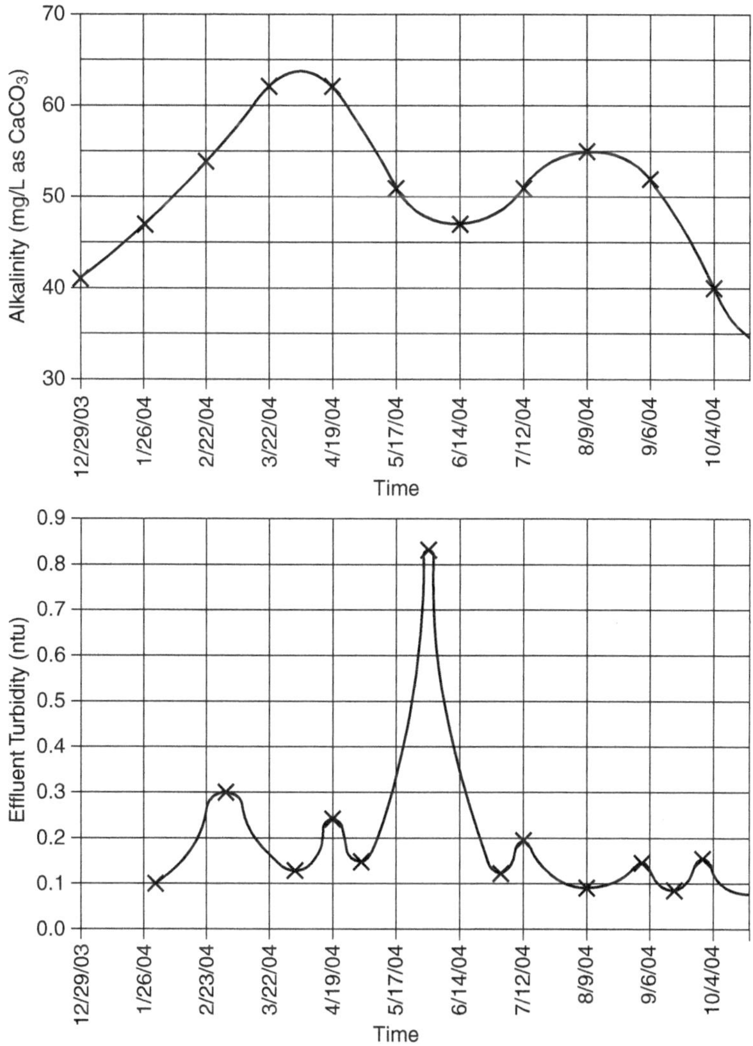

Figure 14-11 Alkalinity and turbidity versus time

NOTE: The turbidity data in the following example were obtained using a standard turbidimeter. In the past, there has not always been a direct correlation between turbidity and particle count data. Using the new laser turbidimeters that are now available may change that relationship. However, laser turbidimeters are not yet in common use in municipal-type treatment plants. Refer also to appendix C.

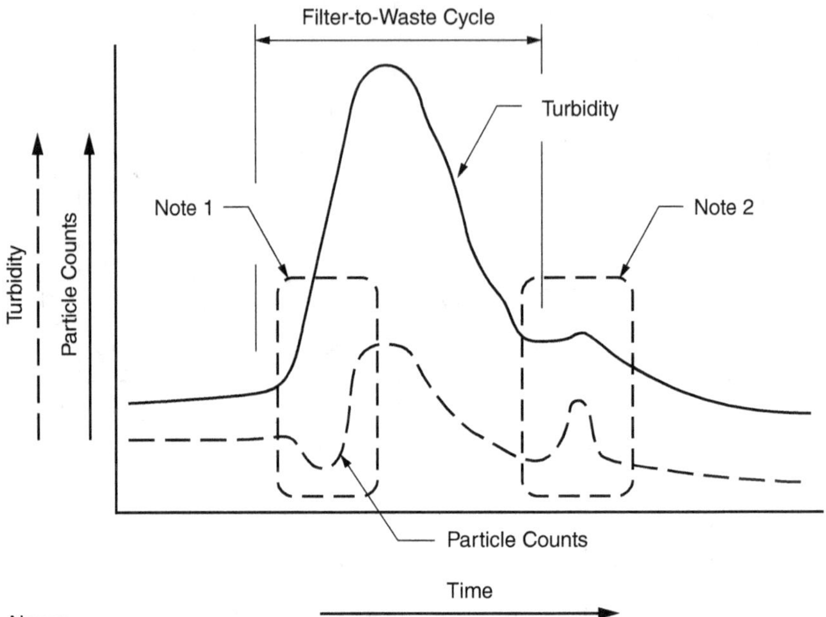

NOTES:

1. Starting up in filter to waste when the filter was idle.
2. Flow surge when beginning filter cycle.

Figure 14-12 Turbidity and particle counts versus time for filter-to-waste cycle

Filter-to-Waste Cycle

An actual graph of particle counts versus turbidity for a filter-to-waste cycle is shown in Figure 14-12. Turbidity is shown by a solid line, and particle counts are shown by a dotted line. An analysis of this graph is a good example of how various different types of data can be used to analyze the function of a treatment plant.

NOTE: For the purposes of this discussion, we will assume that the turbidimeter and particle counter continued to operate throughout the backwash cycle.

Most turbidimeters and other instruments usually do not have automatic valves to close off the sample water supply during backwash. In fact, if a filter is off-line for a long period of time, it is possible for the sample lines to drain the filter. If the filter is to be shut down for a period of time, it would be good to manually shut off all of the sample lines. Then they would have to be opened again when the filter is put back in operation.

Initial stage—steady-state condition. Starting from the left and proceeding to the right on the graph in Figure 14-12, the first stage consists of relatively flat lines. The filter may have been off-line or at the end of a backwash cycle.

NOTE: It should be noted that the turbidity data may have an additional 5-minute delay in response time compared to the particle counts. However, there may also be a delay in the particle counter piping/tubing because of a lower sample flow required. Particle counters operate on a batch basis and can have a response time delay of 30 to 60 seconds.

Second stage—initiate filter to waste. As soon as the filter-to-waste cycle begins, the turbidity begins to rise. However, at the same time, the particle counts go down slightly before beginning to rise again, with no corresponding change in the turbidity. At first glance, the reduction in particle counts at this stage appears to be an anomaly in the data. In order to evaluate this situation, a number of other factors must be identified and considered.

The solids that are rinsed out during a filter-to-waste cycle consist largely of solids that were loosened during the backwash cycle but not completely removed.

When the filter-to-waste cycle begins, the underdrain and effluent/backwash piping are typically full of clean water. Then, when the filter starts again, the initial readings will be of that clean water.

Depending on the depth of the filter, it may take 6 to 10 minutes for the incoming chemically treated water to penetrate down through the media to the sample piping and instruments.

The initial surge of flow, at the beginning of the cycle, is also responsible for a short period of lower water quality.

As stated previously, there is approximately a 5-minute detention time in the body of the turbidimeter, whereas the particle counter recognizes the change in water quality more quickly.

Summary of second stage data. The brief dip in the particle count was probably caused by reading the clean backwash water in the underdrain of the filter, and possibly by fewer larger particles.

The continuous rise in turbidity was most likely caused by the particles being rinsed out during the filter-to-waste cycle, which is the intent of the process.

Other issues may occur in different treatment plants to cause irregularities of this sort, all of which need to be analyzed separately with all the data available. Also, there may be differences of opinion as to what is causing the data variations, especially when similar conditions occur in other plants. These differences are good, in that they show people are thinking. They also allow for more investigation to determine the causes, which is also good for the operators' overall knowledge.

Third stage—turbidity/particle count spike. Both the turbidity and particle count data have a similar spike or peak of elevated values at the same time, although the particle count level is much lower. The spike occurs because of the washing out of loose solids in the media, which is the intent of the filter-to-waste cycle. Once the new chemically treated water penetrates through the media and conditions it properly, the treated water quality begins to improve.

Fourth stage—transition to filter mode. The filter-to-waste cycle may continue for a set period of time or may be terminated when the effluent turbidity reaches a low set-point level. At that point, the filter transitions to normal filter mode.

In the graph on Figure 14-12, another spike of smaller magnitude occurs during the transition to filter mode. The spike in turbidity is very slight, while that of the particle counter is much more pronounced. To evaluate this phenomenon, additional information is required over and above that contained in this figure.

In the actual case, the filter-to-waste flow rate was lower than the normal operational flow rate of the filter. Therefore, when the filter went into filter mode, a flow surge due to the increased rate occurred. The flow surge is responsible for both the spike in turbidity and that of the particle counts. The particle count spike is more pronounced than that of turbidity because the type of turbidimeter used in this case was not as sensitive as the particle counter, and the particles involved are probably much smaller and more numerous.

Again, all the data available should be used to analyze any irregularities.

Filter Runs

Turbidity and particle count graphs for a typical filter run are overlaid on Figure 14-13. On this graph, particle counts are shown with a solid line and turbidity is shown as a dashed line.

At the beginning of this graph, the particle counts are still coming down. Depending on which data are used to terminate filter to waste, the filter run may not be initiated until the particle counts have reached a low set-point level.

Terminal Filter Run Turbidity

Again, depending on which data are used for control purposes, a terminal turbidity set point may not be reached until the far right side of the graph.

Terminal particle count. The decision on when to terminate a filter run and initiate a backwash cycle is site specific, depending of

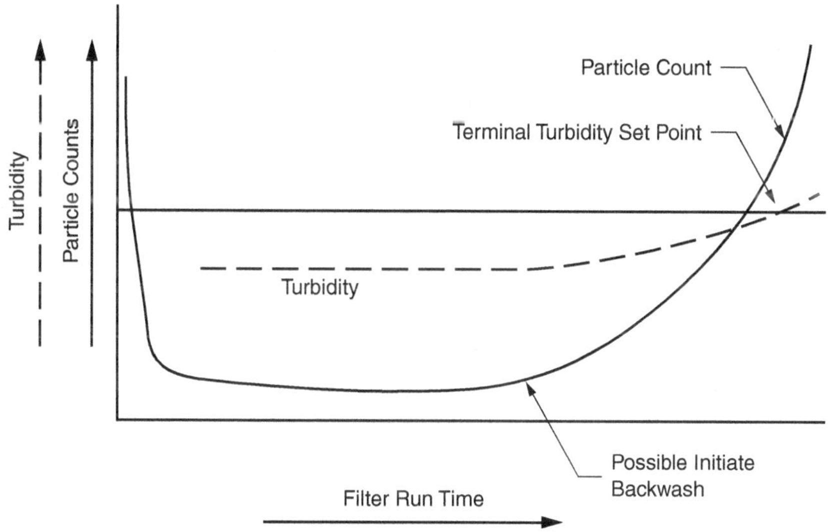

Figure 14-13 Turbidity and particle counts typical filter run

course on the type of data available. If particle counts are being used, the decision to terminate a filter run may be made when the particle counts begin to rise as shown on Figure 14-12. The rise is an early warning that the filter effluent quality is beginning to deteriorate, even though the turbidity is still low. In that case, the backwash cycle may be begun well before a turbidity set point is reached.

Filter run time. Regardless of which data are used to terminate a filter run, it is recommended that the goal for run time still be 24 hours. If it is substantially less, the operational staff should strive toward optimizing the treatment plant and increasing the run time. However, as noted previously, there may be technical reasons or hardware deficiencies that limit the amount of improvement that could be expected.

Particle Count Summary

The level of particle counts to be used to initiate a backwash is typically site specific.

Backwash may be initiated by rising particle counts before the filter's terminal turbidity set point is reached.

It may take longer for particle count readings to stabilize at the end of filter-to-waste cycle than turbidity in some cases.

The response time for particle count data may be considerably less than for turbidity, depending on the size and length of the supply piping/tubing.

SUMMARY

To properly evaluate the operation of a plant, it is necessary to have as much on-line instrumentation as possible. Then, the various data can be compared to help determine the cause of any irregularities.

CHAPTER FIFTEEN

Summary

A great number of issues are involved in evaluating operational data. Unfortunately, smaller plants may not have the manpower required and may rely more on their instrumentation. However, to protect the public health, as much time as possible should be allocated to optimizing the plant, regardless of size.

PLANT SIZE

Many, and perhaps most, treatment plants are of a small size with few operators. Small towns do not normally have the resources to employ many operators. Even so, these smaller plants have the same responsibility and duties as the operators in larger plants. Therefore, some means must be found to insure public health, whether it is more operators, better training, or more sophisticated instrumentation.

ON-LINE INSTRUMENTATION

It is a great benefit to the operators to have all the on-line instrumentation that is pertinent to the operation of the plant, regardless of size. Treatment plants are already required to have effluent turbidimeters. As a minimum, the following additional on-line instrumentation is recommended (not necessarily in this order):

pH/temperature meter (already included in many plants). Three are recommended as a minimum; one on the raw water, one on the chemically treated raw water, and one on the final combined effluent.

A head loss transmitter. One should be provided for each filter and not just a simple switch (also included in many plants, although not on some smaller package plants).

Level probes. One unit should be provided in each filter, one in each compartment of the clearwell, one in each pump well, and one in each storage reservoir.

Streaming current monitor (SCM). One unit should be installed on the chemically treated raw water.

Alkalinity meter. One unit should be installed on the raw water, with another optional unit on the final effluent to help determine the corrosivity of water delivered to the distribution system.

Particle counter. One should be installed on the raw water, one on the effluent of each filter, and another on the combined plant effluent. Although more expensive than most other instrumentation, it is strongly recommended and may eventually be required by the regulatory authorities, if it is not already.

Optional instrumentation might include an SDI meter (if required), and a fluoride analyzer, as well as others that may be required at a particular site. NOTES: Instrumentation does add to the plant expense. However, compared to the overall cost of the plant, the additional cost of the instrumentation is minimal, especially when there are not many operators and most especially because public health is involved. Refer to the flow diagram development in this handbook for a description of instrumentation and the recommended location for each instrument.

MONITORING DATA

All the available data should be reviewed by the operators on a daily basis, as a minimum. As discussed previously, the items of interest are any irregularities and trends in the data that may indicate changes over a period of time.

It should be noted that all irregularities in the data mean something, and there is a reason for them, which needs to be found.

Accumulate Data

All the available data should be accumulated and compared over the same period of time, including weather and raw water data, as well as finished water.

NOTE: As most operators know, one of the most frequent causes of irregularities in a treatment plant is changes in the raw water quality, as long as all the plant equipment is operating properly. These changes can cause the rapid deterioration of finished water quality and should be detected and addressed as soon as possible. For this purpose, SCM data and bench-scale/jar testing are recommended.

Sequence of Events

For any time period being investigated, an accurate sequence of events should be developed including:

1. Record the following status of each of the filters:
    - Filtration mode
    - Backwash cycle

- Filter to waste
- On-line or off-line

2. Record the exact time and sequence of events including:
   - When filters go into backwash and return to service
   - When valves turn on or off
   - When pumps turn on or off
3. Determine the cause of any irregularities
4. Troubleshoot

If no solution is readily apparent, it may be necessary to perform a complete troubleshooting analysis of each piece of equipment per Reference 1.

NOTE: The cause of irregularities can very often be determined by comparing all of the data against the sequence of events at a particular time, or over a particular period of time. While making these comparisons, it should also be obvious that the cause and effect of the various events be known. For additional comments regarding troubleshooting, refer to Reference 1.

PRACTICE CONTINUOUS OPTIMIZATION

Even though the treatment plant may be producing high quality water, the operators should practice continuous optimization, which is to say that they should always try to make the plant run a little bit better than the day before.

- Study the data as continuously as other duties allow.
- Conduct frequent bench-scale/jar testing in an attempt to fine-tune the chemical feed.
- Study each piece of equipment and instrumentation in order to achieve the best possible operation. As an example, if pneumatic valve actuators are used, try to make them operate more smoothly by the use of needle valves. Also, experiment with different settings in PID control loops to determine if there is a better way to make the valves operate more efficiently.
- Perform all required maintenance on instrumentation and equipment. Maintain a maintenance manual with the manufacturers' recommended schedule for this purpose.
- Maintain an operational manual of historical records to help quickly identify process changes that may be necessary. If there is no such manual, write one using all the available data and updating it when other data become available.

Figure 15-1 An out-of-service filter in an active plant

RESPONSE TIME

Identify the process response time of the treatment plant including:

- The process response time to chemical or flow changes.
- Time delays due to lengthy sample piping. Relocate instruments and reduce the size of sample piping if necessary to reduce the response time.
- Identify the response time within the body of various pieces of instrumentation, and account for this time when evaluating and comparing data.

OPERATIONAL MANUAL

A complete operational manual should be prepared that describes the operation of each piece of equipment, as well as the operation of each of the treatment processes. The manual should be complete to the point of providing adequate information for a new operator to be able to run the plant. Operational data should also be provided about the treatment conditions for each typical seasonal variation.

FINAL TEST

If any of the filters in a plant look like the one in Figure 15-1 (an out-of-service filter in an active plant), the process monitoring and evaluation effort is not very good…or even close. A complete troubleshooting procedure needs to be implemented per Reference 1.

REFERENCES

1. Beverly, Richard. 2005. *Filter Troubleshooting and Design Handbook*. Denver, Colo.: American Water Works Association.
2. Beverly, Richard. 2009. *Pump Selection and Troubleshooting Field Guide*. Denver, Colo.: American Water Works Association.

APPENDIX A

Jar Testing Forms and Procedures

JAR TESTING

Make Up Chemical Test Solutions

- Fill one 1,000-mL beaker with distilled water for each chemical used.
- Weight out 1 g of dry chemical or 1 mL of bulk liquid. (More can be used. It just changes the concentrations.) Refer also to the section on chemical calculations for a discussion of dry weight versus bulk liquid calculation.
- Put measured chemical into one of the 1,000-mL beakers of distilled water and stir.
- One mL of the solution will then be 1 mg/L or 1 ppm when put into a 1,000-mL beaker of raw water. (If more than 1 mg or 1 mL is used, the concentration goes up accordingly.)
- Prepare test solutions for other chemicals used in a similar manner (soda ash, polymer, etc.). Please note that the concentration will be different for each.

Equipment Needed

- Jar stirring machine
- 8–10 1,000-mL beakers
- 1-mL pipette
- 5-mL pipette
- 20-mL pipette
- Pipette squeeze bulb
- 100-mL graduated cylinder (or larger)
- Scale for measuring dry powder (if required)
- Distilled water
- pH meter
- Vacuum filter, filter paper, and funnel

Procedure

1. Fill six 1,000-mL beakers with raw water. Make sure to do this quickly before the water warms up. Changes in temperature can affect water chemistry.
2. The primary coagulant should be tested first (alum is assumed herein). Use different amounts of alum in each of the six beakers. Use a uniform spread that brackets the anticipated result; i.e., 2, 4, 6, 8, 10, 12 mg/L. (In some cases, it is desirable to keep one jar as a blank, with no chemical.)
3. Stir rapidly for 30–60 seconds, then slowly for 5–15 minutes, depending on water temperature.
4. Raise the stirring paddles and allow to settle for 10–30 minutes.
5. Observe and record the results. (Poor/good settling, good/fair color, murky, etc.). The proper alum dosage will produce a visible floc that settles, leaving a clear liquid on top.
6. Measure and record the pH of the liquid on the top of each jar.
7. There is normally a fairly narrow pH range for proper coagulation with alum. The pH is normally between 6.3 and 6.8 but can be lower. Raw water with a high pH may coagulate slightly higher but should never be more than 7.0. Above that, aluminum begins to dissolve and pass through the filters.
8. If the coagulation results from step 5 are not good, more alum may be required. If the pH in the beakers drops off rapidly from one jar to the next, lime or soda ash may be required to buffer the water. Buffering will be especially necessary with low alkalinity water typical of coastal regions.
9. If the pH has dropped off, repeat steps 1–6 with the alum dosage and a small mount of lime or soda ash.
10. If the pH range is good and results are not good, add more alum and repeat steps 1–6. If the pH did not drop off before, it may drop off this time. In that case, repeat the test again with the new alum dosages and more lime or soda ash.

NOTES:

a. It may take a number of jar tests to determine the proper dosages. Keep records of all tests. It is recommended that all jar test results be kept in a binder and sorted by raw pH (6.0, 6.1, 6.2…7.0, etc.). A history of your plant will then be developed, allowing quicker changes in the chemical dosages when characteristics of the raw water change. Even if historical records are used, the proper feed rates should be verified by jar tests.

b. Never vary more than one chemical at a time in the same jar test. A systematic approach will be faster in the long run.
c. For waters with high raw pH, it will be necessary to depress the pH by adding more alum, or it may be necessary to switch to a different type of coagulant.
d. For alum coagulation, the coagulation and filtration pH must be in the range of 5–7. Otherwise, soluble aluminum will pass through the plant.

HIGH QUALITY RAW WATER

If the raw water quality is very high, the results from jar tests may not be visible. In that case, a vacuum filter test may be required. The filterability of the water is the ultimate test of the proper chemical feed rates.

- Decant the clear liquid (supernatant) off the top of the beakers.
- Run the sample through a vacuum filter or through folded filter paper and a funnel.
- Measure the turbidity of the filtrate. If it is good, you are through. If not, additional jar tests are required.

POLYMER PROCEDURES

1. Once the proper alum and lime/soda ash dosages have been determined, various amounts of different polymers can be tried. Use the proper amount of alum and lime/soda ash in each jar and vary the amount of polymer.
2. Polymer dosages should not normally exceed 2 mg/L on a bulk liquid basis. For dry polymer, the dosage should probably not exceed 0.1–0.2 mg/L on a dry weight basis. Verify the maximum allowable dosages with the manufacturer.
3. *Caution:* Do not overfeed polymers. An excess can glue the filter media together.
4. Refer also the manufacturer's recommendations for maximum allowable feed rates. Refer to the following section on "Chemical Calculations" for a discussion fo dry weight versus bulk liquid calculations.

CHEMICAL CALCULATIONS

Once the desired chemical feed rates have been determined by jar testing, it is necessary to convert those values to pumping rates.

Calculate one part per million (mg/L) based on a full day's production, even if the plant does not run that long. There are two methods of making calculations depending on the type of chemical used, either by using dry weight or on a bulk liquid basis.

Bulk Liquid Alum

$$1 \text{ mg/L} = \frac{\text{plant flow (gpm)} \times 1{,}440 \text{ (min/day)}}{1{,}000{,}000} = \_\_\_ \text{ gpd bulk liquid alum}$$

Please note that the actual concentration of the bulk liquid does not matter. If the bulk liquid used for jar tests is the same as the plant uses, the concentration cancels out in the calculations and it results in simpler computations.

Dry Alum, Powdered Polymer, Lime or Soda Ash

$$1 \text{ mg/L} = \frac{\text{flow (gpm)} \times 1{,}440 \text{ (min/day)}}{1{,}000{,}000 \times \_\_\_ \text{ dry chemical concentration (lb/lb of water)}} = \_\_\_ \text{ gpd mixed chemical}$$

For example, assuming a plant flow of 700 gpm and a soda ash mix concentration of ½ lb to 1 gal water (6%) would yield the following results:

$$1 \text{ mg/L soda ash} = \frac{700 \text{ gpm} \times 1{,}440 \text{ min/day}}{1{,}000{,}000 \times 0.06 \text{ mix concentration}}$$

$$= \frac{1 \text{ mgd}}{1{,}000{,}000 \times 0.06}$$

$$1 \text{ mg/L soda ash} = 16.67 \text{ gpd}$$

Calculate Chemical Flow

- Assume a plant flow of 700 gpm (1 mgd)
- Assume bulk liquid alum
- mg/L from jar test—assume 8 mg/L

- 1 mg/L bulk liquid alum = 1 mgd/1,000,000
  = 1 gpd bulk liquid alum
- Daily chemical flow = 8 ppm × 1 gpd (jar test)
  = 8 gpd

Calculate Alum Pump Setting

- Assume 24-gpd pump
- Pump setting $= \dfrac{\text{daily chemical flow (gpd)}}{\text{pump capacity (gpd)}} = \dfrac{8}{24} \times 100 = 33\%$
- NOTE: If the pump has a speed and stroke setting, they must be multiplied together to give the above result.
- Pump setting (%) = speed × stroke

  = 57% × 57% $_8$0.33

  = 33% overall

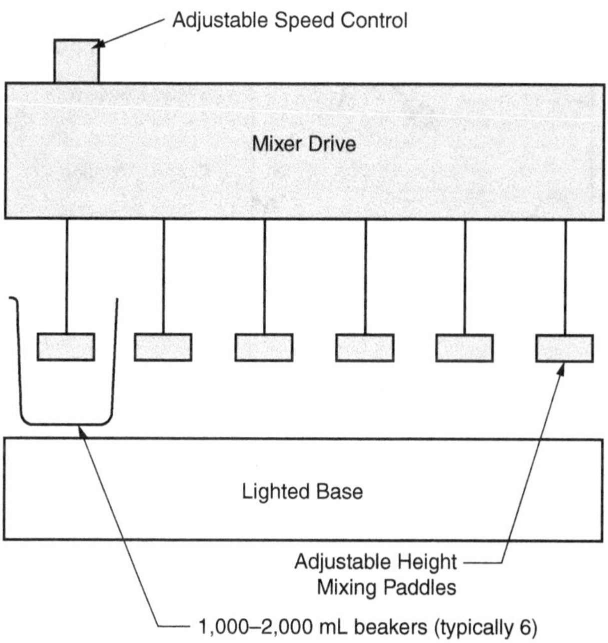

Figure A-1 Typical jar testing machine

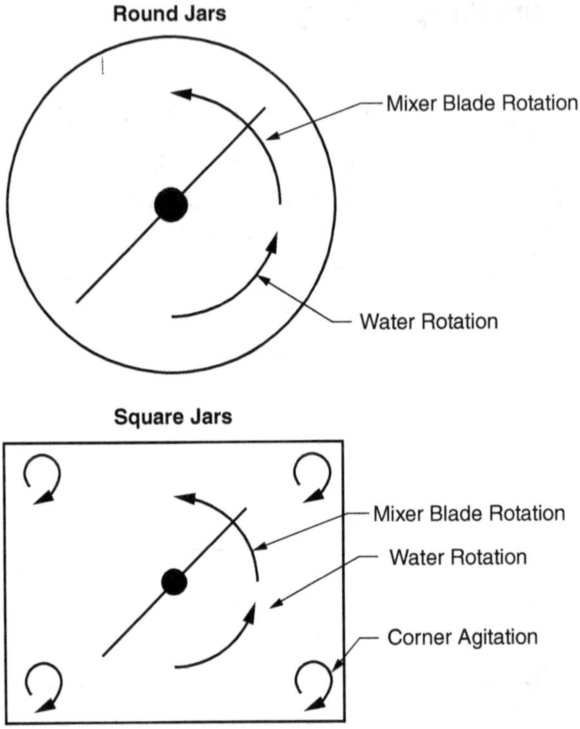

Note: Mixing is more efficient with square jars due to corner agitation.

Figure A-2 Jar types

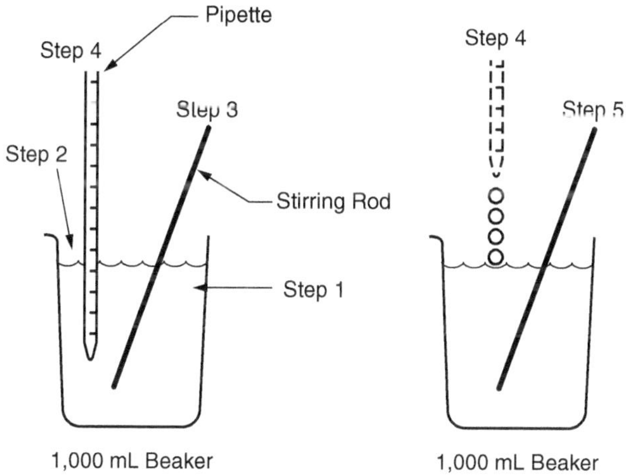

Step 1	Fill both beakers with distilled water.
Step 2	Add 1 g dry chemical or 1 mL of bulk liquid.
Step 3	Stir first beaker thoroughly.
Step 4	Take 1 mL of mixed solution and pour into second beaker.
Step 5	Stir second beaker thoroughly.
Step 6	1 mL of second beaker will now be equivalent to 1 ppm when put into 1,000 mL of raw water.

Figure A-3 Chemical dilution procedure

______________ Water System

Jar Testing Log — Time ________

Test No. ________ — Date ________

Raw pH ____________
Raw Alkalinity ____________
Raw Turbidity ____________

Jar No.	1	2	3	4	5	6
Alum (mg/L)						
Lime (mg/L)						
Soda Ash (mg/L)						
Potassium Permanganate (mg/L)						
Other (mg/L)						
Polymer (mg/L)						
Type ____________						
Polymer (mg/L)						
Type ____________						
Treated pH						
Results						

NOTES:

Alum ☐ Bulk liquid basis ☐ Dry powder basis ☐ Dry granular basis
Polymer A ☐ Bulk liquid basis ☐ Dry powder basis
Polymer B ☐ Bulk liquid basis ☐ Dry powder basis

Figure A-4 Typical jar testing form

APPENDIX B

HISTORICAL RECORDS

Date: __________

I. Plant Flow Rate __________ gpm or mgd

II. Raw water quality

Turbidity	____ ntu
pH	____
Alkalinity	____ mg/L
Water Temperature	____ °F (°C)
Color	____
Suspended Solids	____ mg/L
Silt Density Index	____
Etc.	____

III. Chemical Feed Rates

Chemicals	Rates	Pump Settings (Stroke/Speed)
Polymer 1	____ mg/L	___________
Polymer 2	____ mg/L	___________
Coagulant	____ mg/L	___________
Powdered Carbon	____ mg/L	___________
Potassium Permanganate	____ mg/L	___________
Soda Ash	____ mg/L	___________
Lime	____ mg/L	___________
Sodium Hydroxide	____ mg/L	___________
Chlorine	____ mg/L	___________
Other	____	___________

IV. SDC Calibration/Set Point __________

V. Clarifier/Sedimentation Basin Effluent __________ ntu

VI. Filter Effluent Turbidity __________ ntu

VII. Run Time Between Backwashes __________ hr

VIII. Filter Effluent Particle Count (optional) __________

IX. Clean Bed Head Loss __________ ft

X. Terminal Head Loss __________ ft

XI. Filter Effluent

Turbidity	____	ntu
Chlorine Residual	____	mg/L
pH	____	
Silt Density Index	____	

Note: Other data may be added as required for a specific site.

APPENDIX C

Treatment

Measuring Turbidity in the Laser Age

Detecting early filter deterioration can be tricky. Laser nephelometers can help by providing accurate on-line measurements to help facilities meet regulatory reporting requirements and optimize filter performance and run times.

BY MICHAEL SADAR, STEVE CASON, AND TERRY ENGELHARDT

CONVENTIONAL WATER filtration plants are usually required to monitor turbidity levels in filter effluent. Until recently, the lowest level of change that turbidimeters could measure was too high to detect early filter breakthrough, so many water treatment plants turned to particle counters. However, turbidimeter technology advancements now include laser nephelometers whose detection capabilities approach the functionality of particle counters.

Laser nephelometers can detect ultralow turbidity changes, as well as filter deterioration and breakthrough. Increasingly, these capabilities allow turbidity and particle-counting measurements to become complementary technologies. In addition, a laser nephelometer plays two roles at conventional water treatment facilities by providing accurate, on-line turbidity measurements to meet regulatory reporting requirements and serving as a sensitive, highly accurate process instrument for optimizing filter performance and run times.

FEATURES AND BENEFITS

Laser nephelometers use advanced incident light to reveal a sample's scattered light reading (Figure 1). A 35-mW solid-state laser diode, rather than incandescent light, introduces a high level of sensitivity to nephelometry. Sophisticated laser optics and signal processing allow the laser nephelometer to detect turbidity changes as low as 0.0003 ntu, as well as submicron-size particles (smaller than .01 µm), which typically precede larger particles.

Laser nephelometer use is becoming common at membrane-based drinking water treatment plants in which operators usually are looking for evidence of clogged membranes and broken membrane fibers, primary issues with subtle early warning signs. Membrane plant operators monitor for evidence of slight pressure shifts and small changes in turbidity, miniscule changes that can be regularly detected with a laser nephelometer. Conventional visible light (US Environmental Protection Agency method 180.1) or long-wavelength light-emitting diode-based nephelometers (ISO7027) aren't sufficiently sensitive to detect minute changes in membrane systems.

In contrast, conventional water filtration plants monitor filter effluent for any protracted rise in turbidity. These plants have long used particle counters to detect extremely small changes in performance that standard turbidimeters

Figure 1. Laser Nephelometer Optical Configuration

Laser optics and signal processing allow a laser nephelometer to detect turbidity changes as low as 0.0003 ntu.

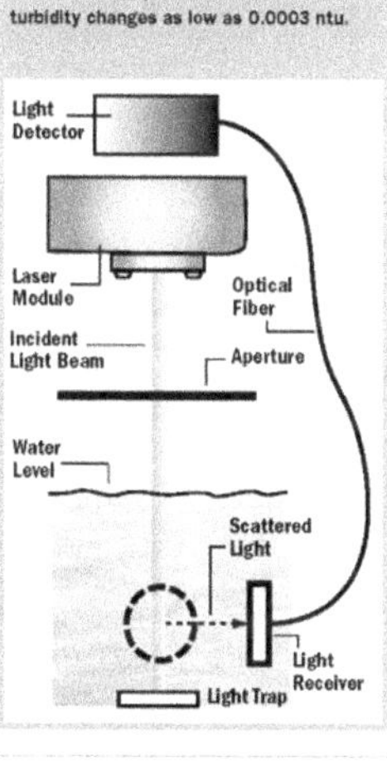

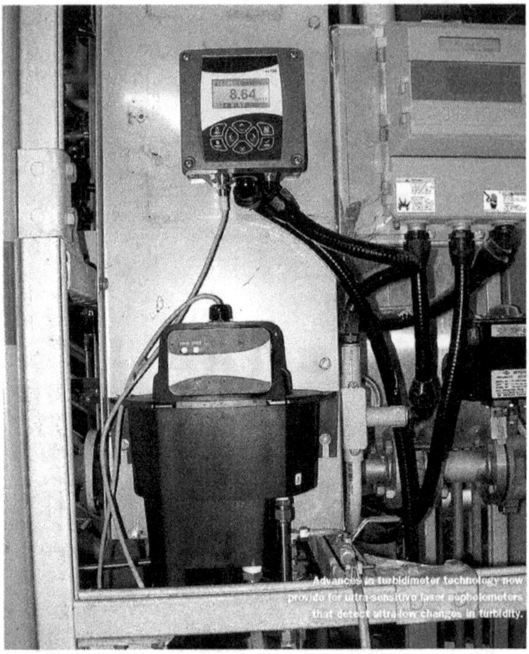

Advances in turbidimeter technology now provide for ultra-sensitive laser nephelometers that detect ultra-low changes in turbidity.

PHOTOGRAPH: HACH

Michael Sadar, Steve Cason, and Terry Engelhardt are with Hach (www.hach.com), Loveland, Colo.

can't detect. However, long-term, successful application of laser nephelometers in membrane plants to detect small step changes in the filtering process indicates this instrument also can effectively monitor and optimize conventional granular media filter performance.

SUBMICRON PARTICLE EVENTS

A data-collection study was conducted at a conventional 30-mgd, 12-filter water treatment plant near Fort Collins, Colo., to determine, among other things, effective technologies for turbidity spike detection in filter effluent. Three types of instruments were used to monitor plant effluent:

- A particle counter, with size sensitivity down to 2 µm.
- A low-level regulatory-approved turbidimeter commonly used in conventional water treatment plants for regulatory filter effluent monitoring, which meets all instrument-design criteria specified by USEPA method 180.1.
- Three laser nephelometers that are about 150 times more sensitive than traditional turbidimeters and can confirm particle events that might otherwise be reported on traditional low-level turbidimeters as noise. The instruments measure turbidity in mntu units where 1 mntu = 0.001 ntu.

All instruments were used alongside regular sampling. Redundant testing with three laser nephelometers was also performed to detect minor events and isolate interferences such as bubbles or contamination.

To test its effectiveness on a conventional filter, the laser nephelometers were operated with the particle counter. Turbidity spikes were confirmed as detected by a laser nephelometer only if all three laser nephelometers detected the spikes. Particle events detected by all instruments were examined and data were collected during 66 continuous filter runs. Although the laser nephelometer and particle counter produced mostly complementary data, the laser nephelometer's sensitivity allowed it to detect submicron particle events that were otherwise undetectable (Figure 2).

Using a particle counter and a laser nephelometer also provided information about the composition of a particle event. Events detected by both instruments indicated a natural distribution of particles roughly following a $1/d^3$ relationship. In these cases, a nephelometer will detect particles slightly before a particle counter because small particles move more rapidly through the filter (Figure 3).

ASSESSING FILTRATION EXCURSIONS

The laser nephelometer automatically calculated and displayed relative standard deviation (RSD), which provided even greater sensitivity to pending filtration breakthrough or loss of membrane integrity. A separate, distinct, and dimensionless monitoring parameter, RSD quantitatively assesses turbidity fluctuation. RSD and turbidity measurement values are complementary and should be recorded and trended using the laser nephelometer's analog or digital output capabilities. Careful observation of the relationship between the two values provides an excellent diagnosis of the process excursion.

RSD is calculated as the standard deviation for a given set of measurements divided by the average for the same set of measurements. The quotient is then multiplied by 100 to express the result as a percent. For example:

$$\text{RSD} = (\text{Stdev}_n / \text{Average}_n) \times 100$$

Here n equals a defined number of measurements used to calculate the average and the standard deviation.

RSD can be used to monitor pending filtration breakthrough on conventional granular media filters because RSD will increase prior to any marked turbidity level increases. The parameter also enhances the sensitivity of detection for membrane integrity loss for microfiltration, ultrafiltration, nanofiltration, and reverse osmosis. Combining RSD and

Treatment

Figure 2. Detailing a Particle Event

During a filter run, the turbidimeters detected all particle events. The particle counter missed an event at approximately 14:24.

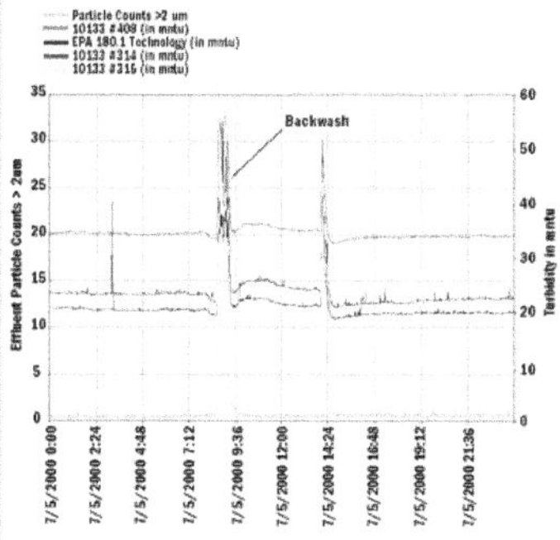

Figure 3. Measuring Sensitivity

Laser turbidimeter baselines increase in fluctuation, indicating the bleeding of low numbers of larger particles through the filter.

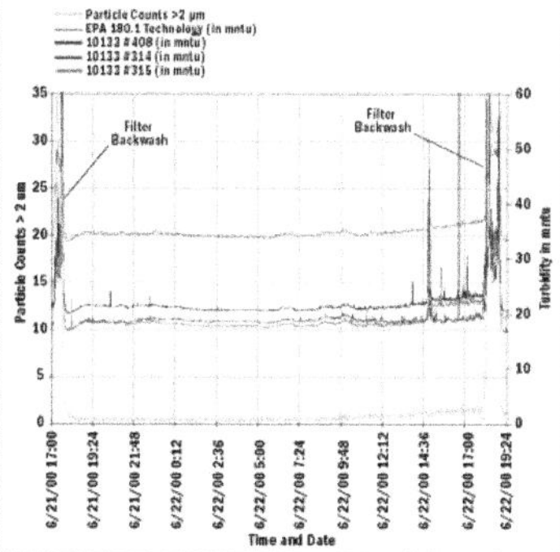

laser turbidity readings can provide a qualitative assessment regarding the nature of membrane integrity loss.

FIELD TEST

The laser nephelometer was also evaluated at a conventional 160-mgd water treatment plant in Southern California in summer 2007. The quality of the plant's raw water from the American River is usually high, with influent measuring below 2 ntu in the spring, with rare spikes of more than 10 ntu when the area experiences storm runoff. The plant runs 16 full-media sand and anthracite gravity filters consisting of eight filters built around 1964 and eight filters added in 2005.

The plant is not required to count particles but has been doing so to detect early breakthrough. A field test of the laser nephelometer ran in conjunction with a two-week flow test the plant underwent as part of an independent evaluation to determine component performance under high hydraulic loadings.

The plant's flow test provided a suitable testing ground for the laser nephelometer because filter breakthrough events are more likely when a plant pushes hydraulic processes. During the two-week flow test, a laser nephelometer was added to one of the filters already served by a particle counter and turbidimeter. During this test, data showed that the laser nephelometer detected the same breakthrough events as the particle counter. Readings from the conventional turbidimeter, however, remained flat. The laser nephelometer is easily calibrated and integrated well with the plant's supervisory control and data acquisition system.

VERIFICATION

Regulations are placing an increased emphasis on verifying on-line instrumentation. The laser nephelometer used in these field studies features a solid, highly stable immersion standard that allows calibration to be checked intermittently between wet calibrations.

The 20-second, self-indexing feature verifies calibration below 0.1 ntu (100 mntu) with no sample cells to recondition to bring the instrument back online. Detecting impending filter breakthrough, delineating filter ripening, maximizing effective filter run times, and collecting regulatory reporting data with the same instrument isn't new to membrane plant operators who rely on the sensitivity of laser nephelometers. These capabilities are also available to conventional water treatment plant operators as an alternative to particle counters.

Conventional water treatment plants are undergoing change as requirements gradually become more stringent. In this shifting regulatory environment, conventional water treatment plants can benefit from turbidity monitoring technology that detects ultra-low turbidity levels and submicron particles.

APPENDIX D

Technical Bulletin TB-PCNT1

Particle Counting in Drinking Water

Definition and Measuring Principle

Particle Counters use one of three different sensing technologies to count and size particulate matter in a liquid process:

1. **Light Scattering (Figure 1)**—This technology is similar to the method employed by modern turbidimeters. Light is directed through a process and, as light is reflected off the surfaces of particles within the process, it is measured at an angle divergent from its original path. The amount of light received at the detector is proportional to the size and the index of refraction of the particle being measured.

Figure 1 Light scattering technology

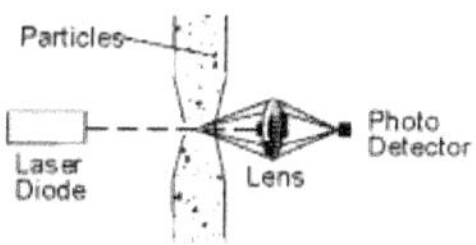

2. **Electrical Resistance (Figure 2)**—Historically, this technology has been used in the medical field. This method requires the measured liquid to be conductive. An electrical current is measured across a small aperture using two electrodes. An electrical pulse is generated whenever a particle passes through the aperture. The pulse is proportional to the volume of the particle.

Figure 2 Electrical resistance technology

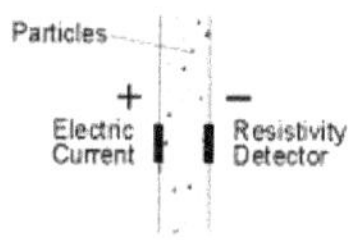

3. **Light Extinction (Figure 3)**—The most common technology used for counting and sizing particles in drinking water is light extinction or light obscuration. In this method, a sample is passed through a small chamber, usually constructed of sapphire or glass. A light source, normally a high intensity laser diode, is focused through this sample cell. As a sample flows through the chamber, particles within that sample absorb or scatter some of the light that illuminates them. A photo detector is positioned across the cell from the laser. It measures the amount of light received and the amount of light blocked by the particle passing through. The amount of light blocked is proportional to the size and the index of refraction of the particle being measured.

Figure 3 Light extinction technology

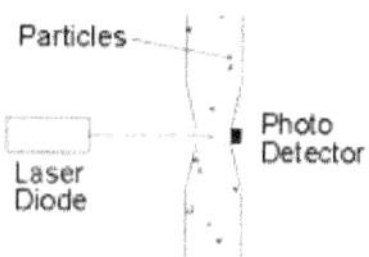

For on-line particle counters, the flow rate through the cell is of utmost importance. Total counts are related to the volume of water being sampled (for example, 20 counts per milliliter).

Technical Bulletin TB-PCNT1

Insitu versus Volumetric Measurement

There are two kinds of light extinction particle counters. Volumetric particle counters use lasers focused across the entire flow cell illuminating a cross section of flow. These units theoretically count all particles that pass through the cell. They are, however, prone to measurement error due to air bubbles and buildup on the cell walls that interfere with light transmission. Volumetric particle counters typically have smaller flow cells, which are also more prone to clogging in certain applications such as raw water monitoring.

Insitu particle counters use a highly focused beam of light to illuminate a portion of the flow cell, and then extrapolate the concentration of particles from that information. The advantages of insitu particle counters are:

- Less interference from air bubbles
- Lower coincidence measurement error
- Sample flow is less susceptible to blockage because of an inherently larger flow cell.

In actuality, all particle counters are insitu in one critical way. Since no currently available system is capable of measuring all particles in all water leaving a treatment plant, only a small portion of the total effluent gets measured. For example, if a plant produces 15 MGD and has 10 particle counters, only 50 to 60 gallons will pass through the particle counters per day, no matter which method is used. That is why proper placement of these devices, and adherence to manufacturers' guidelines is critical.

Measurement Discrepancies between Particle Counters

It is highly unlikely that two particle counters measuring the same sample will yield the exact same results. This is due to several factors, most importantly the index of refraction of the particles being measured. As a particle moves through the sample flow cell, different faces are exposed to the light. As with any three-dimensional object of irregular shape, some surfaces reflect light more than other surfaces. The greater the blockage of light, the larger the device perceives the particle to be.

With some insitu systems, it is possible to do a count match in the field. By using one unit as a master, adjustments through the software can allow instruments to read the same. With Hach's AccuCount+ particle counting system, this is simply done by adjusting the percent viewable volume of the laser in the diagnostic window of Hach's optional AccuView software package.

Coincidence Measurement Error

Coincidence is the measurement of multiple particles at the same time, which the particle counter perceives as one discrete particle. A large particle can mask smaller particles between it and the photo-detector, or multiple particles can chain together to form what appears to be one long particle. These measurements both result in error known as coincidence. Properly sizing the flow cell can minimize the errors associated with coincidence. Another important factor in controlling this undercounting of particles is properly adjusting the flow rate through the sample flow cell. Therefore, volumetric particle counters, which have a larger laser footprint, are more susceptible to coincidence measurement error than insitu devices.

Maintenance

Like all on-line instruments, particle counters require some regular maintenance. The intervals between maintenance are dependent upon the operating environment of the instrument. If a system is used to detect particles in raw water entering a filtration plant, and the water has very high particle counts, a weekly or even daily cleaning of the plumbing from the flow controlling device to the sensor may be needed. Also a thorough cleaning of the sample flow cell may be required. In areas without the high quantities of particulate matter, such as post filtration, the need for maintenance may be as minor as changing the desiccant cartridge on a quarterly basis.

Calibration

New particle counting systems are factory-calibrated by the manufacturer. Unlike many other on-line instruments, calibrating particle counters is a time-consuming process requiring the use of specialized equipment. For this reason, only the manufacturer or a qualified technician should perform calibration. Typically, the calibration of each particle counter should be verified annually. Based on results of field verification procedures, it may be re-calibrated to ensure best particle counting accuracy.

Page 2
Particle Counting in Drinking Water

Field Procedure for Verification of Calibration

Re-calibration is required when optics become misaligned, light output from the laser changes, or optics become damaged or dirty. Verification is the easiest way to determine if a particle counter needs to be serviced. Actual counts are difficult to verify in the field. It should be assumed that particle sizing represents a surrogate for actual particle counting accuracy.

Field Procedure for Verification of Calibration

Thoroughly clean and fill two one-gallon containers with sub-micron filtered (distilled) water. Place one container higher than the top of the flow controlling device. Remove the sample inlet hose from the flow controlling device, and replace it with a short section of similar-type hose. Place the other end of the short section hose in the one-gallon container, with its end close to the bottom. Induce a siphon through this hose to the flow controlling device by placing the one-gallon container lower than the flow controlling device. Water from the flow controlling device will fill the hose, forcing out all the air. Now raise the container higher than the flow controlling device and a siphon should be established.

With the sub-micron filtered (distilled) water flowing through the instrument, note the count data. It will decrease after a minute or two as the resident process water is flushed out. Wait for counts to stabilize. This may take nearly the entire gallon of filtered water. If it becomes necessary, refill the container with more sub-micron filtered (distilled) water.

While waiting for counts to stabilize, thoroughly mix a small amount (about 2 ml) of standard (mono-dispersed polystyrene spheres) into the other gallon of sub-micron filtered (distilled) water. Thoroughly mix by gently inverting the container about 25 times.

> **Note:** *If the solution is mixed too vigorously, air bubbles may become entrained. Also, by adding too much standard to the suspension, particle counts may appear as larger sizes due to coincidence measurement error.*

Switch the inlet hose from the first container to the second one containing the particle suspension. If necessary, re-establish the siphon. Wait for 5 minutes. You should observe increased counts in the channel(s) corresponding to the size standard mixed in the water.

Depending on the sampling interval of the particle counter, it may be necessary to wait several more minutes for counts to stabilize.

> **Note:** *If multiple bead sizes will be used, start with the largest particles; they clean out quicker and are present in lower quantities than smaller particles.*

After the standard has been introduced to the particle counter and the counts have stabilized, observe its output. A properly calibrated unit will show a count increase in the channel corresponding to the size of the standard used (for example, more counts in the 10-15 micron channel when a 12 micron standard is used).

An improperly calibrated unit will show either no increase in any channel, or increased counts in a channel other than the one corresponding to the size of the standard. An exception to this is when the size of the standard falls on the sizing threshold of two adjacent particle size channels. Example: Suppose 5 micron beads are used and the adjacent sizing channels are 2-5 micron and 5-10 micron. In this case, increased counts should be observed in both channels. If increased counts are only observed in one of the channels, or there is no count increase, a re-calibration should be performed.

Figure 4 **Typical drinking water particle counting system**

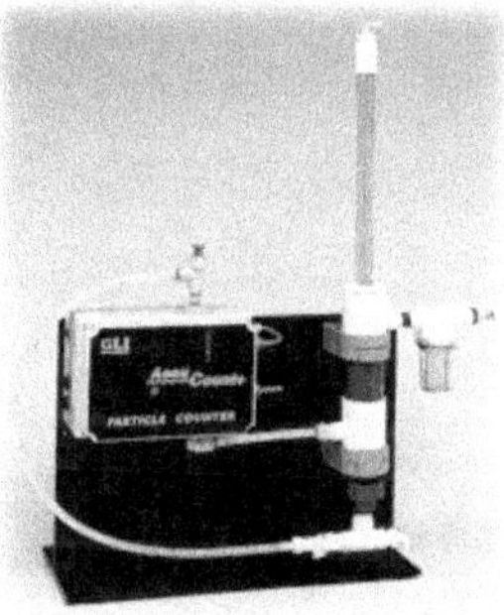

Technical Bulletin TB-PCNT1

General Installation Guidelines

Installation is critical to a particle counter's ability to accurately assess water quality. By gathering the sample incorrectly, an erroneous number of particles can be measured. Interference such as air bubbles or particles sloughing off the inner walls of a pipe can falsely indicate filter problems that may not actually exist. A proper sample line installation should enable a smooth change of direction of the sample within the main process pipe. This reduces the chances of off-gassing in the sample line. The particle counter should be installed as close as possible to the point of the desired measurement. By minimizing the tubing length, there is less chance of particles settling out or impurities from the line being picked up on the way to the sensor. It is also critical that samples for other particle counters throughout the plant be gathered using similar installation practices. This allows for more accurate comparisons between instruments for functions such as log reduction and filter efficiency calculations.

Conclusion

Particle counting can provide very useful data for monitoring water quality in modern drinking water treatment plants. The data can be used to diagnose filter performance, evaluate changes to plant operation such as increased alum feed rates, and determine filter efficiencies. When used with the appropriate software, data can be collected and maintained, enabling operators to properly document facility performance. At this time there are no federal regulations requiring the monitoring of particle concentrations in drinking water. However, the real benefit of gathering particle count data lies in the user's ability to fine-tune filter performance and assure that the best possible water is being produced.

FOR TECHNICAL ASSISTANCE, PRICE INFORMATION AND ORDERING:
In the U.S.A. – **Call toll-free 800-227-4224**
Outside the U.S.A. – **Contact the Hach office or distributor serving you.**
On the Worldwide Web – **www.hach.com**; E-mail – **techhelp@hach.com**

HACH COMPANY
WORLD HEADQUARTERS
Telephone: (970) 669-3050
FAX: (970) 669-2932

APPENDIX E

Application Note 120

A Strategy for Optimizing Water Treatment Plant Performance Using Light Scatter Technologies

Executive Summary

Two new particle detecting technologies have been developed to help optimize filter performance at water treatment plants (WTP). One nephelometer was designed to give simple, accurate, and rapid response to turbidity changes during a backwash cycle and while monitoring the filter effluent. The other nephelometer was designed specifically to monitor filter effluent. During this study, the instrumentation was primarily used to monitor particle events during the WTP process. The ultimate goal is to optimize plant performance by identifying and reducing particle events that occur either before or after final filtration.

Optimization of the filter run was defined as the production of a stable effluent stream (characterized by low and consistent turbidity and low and consistent particle counts). Particle shedding from the filter into the sample was minimal for the duration of the run.

Two new instruments were used in this study: a laser nephelometer and a probe turbidimeter. The laser nephelometer, which is designed to detect very small changes in turbidity, was combined with a particle counter and regulatory turbidimeter on the filter effluent. The probe turbidimeter exhibits quick response and contains an 860-nm infrared light source, making it immune to color interference. The probe was positioned in the influent immediately above the filter. Collectively, this instrument distribution allows for more in-depth profiling of each particle event as it moves through the filter.

The study involved the participation of a local water treatment plant that is a member of the Partnership for Safe Drinking Water. For the past year, this plant has been heavily involved in the development and testing of this laser nephelometer. The data collected from that testing was used as a baseline for comparison to data generated in the study.

Site Profile

The water treatment plant where this study was conducted is located near Fort Collins, Colorado. A member of the Partnership for Safe Drinking Water, the plant has the capacity to produce 30 MGD using 12 filters. For the purpose of this study, a single filter was evaluated. The plant's current goal is to not exceed 0.1 NTU in the effluent, even during a backwash event. This plant's processes are under excellent control but the management and operators are interested in continuous improvement by further optimizing their filter runs.

During this study, the plant undertook an expansion project to increase its production capability to 50 MGD. Also, the raw water source, a reservoir, experienced significant change due to severe drawdown (draining of the reservoir). The geographical area supplied by the plant experienced significant drought conditions that required the plant to run at or near capacity for the duration of the study.

Application Note 120

Introduction

Particle events are often viewed as surrogates for the quality of water produced in a WTP. Fewer events indicate a higher filter performance and therefore, better water quality. In this study, particle events detected by either a particle counter, laser nephelometer, or standard regulatory turbidimeter in the effluent stream were examined. Monitoring for particle spikes was performed at two points prior to filtration. A standard turbidimeter (1720C) was used to monitor the water as it exited the sedimentation basin and an OptiQuant™ SST probe turbidimeter was used to monitor the water just before passing through the filter.

The instruments were strategically positioned in the treatment stream to help determine if spikes that were detected leaving the sedimentation basin travel through the filter. If they did travel through the filter, the goal was to determine if the spikes changed before and after filtration. The magnitude and duration of each spike was also analyzed at different phases of the treatment process.

Laser nephelometers, regulatory turbidimeters, and a particle counter were used to monitor the filter effluent in an effort to determine if the instruments are complementary (which instruments identify the same particle event) or if they detect different events. This comparison of instruments provides WTP management insight into which instrument technology will help them optimize their filtration management. In addition, overall plant performance during plant expansion and geographical drought can be evaluated.

During this study, process monitoring was conducted for a total of 66 continuous filter runs on a single filter. The goal was to focus on the collection, preparation, and analysis of the data without impacting the day-to-day plant operation. All monitoring was passive and the data was analyzed after collection.

Four primary goals were set for this study:

1. To evaluate the role of different technologies in filter optimization and continuous improvement in this DWP.
2. To provide more insight into the WTP processes and the impact, if any, on particle events as they move through a water treatment process.
3. To determine which technologies are better suited for the detection of particle spikes before and after the filter.
4. To investigate the relationship between influent spikes and final effluent turbidity and particle counts.

Materials and Methods

Instrumentation
Three types of instruments were used for effluent monitoring event detection: a particle counter, a low-level regulatory-approved turbidimeter, and three laser nephelometers. All instruments monitoring the effluent were run in parallel with regular sampling.

- The 1900 WPC Particle Counter used in the study has size sensitivity down to 2 microns. For consistent and reliable application of the instrument, it was positioned on the effluent side of the filter.

Application Note 120

- The regulatory turbidimeter was a Hach 1720D. The 1720D is commonly used in WTPs for regulatory filter effluent monitoring. This instrument meets all instrument design criteria specified by the USEPA method 180.1.
- FilterTrak™ 660 Laser Nephelometers were also used. These instruments are approximately 150 times more sensitive than traditional turbidimeters and will confirm particle events that might otherwise be interpreted as noise on a traditional low-level turbidimeter. The FilterTrak 660 measures turbidity in mNTU units (where 1 mNTU = 0.001 NTU).

Above the filter, two types of instruments were used:

- A Hach 1720C turbidimeter, owned by the WTP, monitored the sample as it left the settling basin.
- A new turbidimeter, the OptiQuant SST, was installed on the settled water immediately above the filter. This probe design instrument utilizes ISO method 7027 design criteria for turbidity monitoring. Characterized by its quick response, the probe turbidimeter is often used for profiling events, including the turbidity of backwashes.

Table 1 summarizes the instrumentation used in the study. Figure 1 shows the strategic location of the instruments in this study.

All instruments were polled simultaneously at 1-minute intervals and data were logged to a computer using digital data networking protocol to minimize errors in measurement and transcription. Microsoft® Excel® was used to analyze and graph the data.

Table 1 Instrumentation Used in the Study

Location	Instrument	Primary Application
Filter Effluent	1900WPC Particle Counter	Counts and profiles particles that are >2µm in size
Filter Effluent	1720D Turbidimeter	Regulatory low-level turbidity
Filter Effluent	FilterTrak 660 SN 408	Low-level spike detection
	FilterTrak 660 SN 314	
	FilterTrak 660 SN 315	
Settled Water	1720C Turbidimeter	Turbidity in the 0.5–5 NTU Range
Applied to Filter	OptiQuant SST Turbidimeter	Fast Response in 0.3–1000 NTU Range

Redundant testing using three FilterTrak 660 Nephelometers was performed to increase confidence in the new technology, to confirm the detection of minor events, and to isolate interferences such as bubbles or contamination.

Once installed, all instrumentation was calibrated according to the manufacturer's instructions. After calibration, the instruments were allowed to run continuously from May 6 to July 15, 2000 until the 66 filter runs were completed. At approximately four-day intervals, the data was downloaded and analyzed for particle events and other significant criteria.

Application Note 120

Figure 1 Instrument Location in the Water Treatment Plant

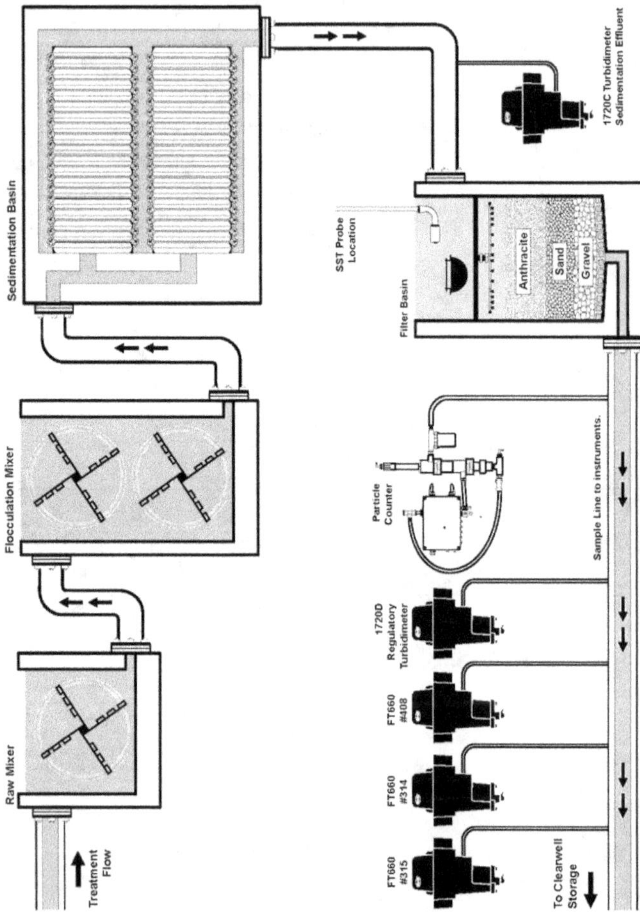

Application Note 120

Particle Events:
Events are characterized as either major or minor. For this study, a major event is categorized as a turbidity spike that is greater than 5 mNTU and that lasts longer than 5 minutes. For particle counting, a major event is any sustained count spike that is greater than 2 counts per mL. Using this criteria ensures that bubbles are not identified as events. A turbidity minor event is any spike that is between 1 and 5 mNTU, or any change between 1–2 counts per mL above the baseline on a particle counter. These criteria only apply to the filter effluent. The events are summarized in Table 2.

Table 2 Particle Event Characterization in Filter Effluent

Instrument	Major Event	Minor Event
FilterTrak 660 or 1720D	>5 mNTU above baseline	1–5 mNTU above baseline
1900WPC Particle Counter	>2 cts/mL above baseline	1–2 cts/mL above baseline

Note: Natural particle distribution follows a $1/d^3$ relationship with respect to number and size. Each order of magnitude decrease in particle size shows approximately 10^3 more smaller particles.

Depending on which instrument detects the event, the event profile can be determined. If the event is seen only by the turbidimeter and not by the particle counter, the particles are assumed to be sub-micron. If the event is seen only by the particle counter, then the particles of that event are greater than 2 µm in size and exist in very low numbers. Events that are observed on both instruments indicate that the spike contains a natural size distribution after passing through the filter. An example of this distinction is shown in the data section.

Event detection using laser nephelometer technology required simultaneous detection of the particle spike by all three FilterTrak 660s in the study. When all three instruments detected the spike, the presence of the spike was confirmed.

Data

Once the data was collected and plotted, several pieces of information (metrics) were entered into a master matrix. These include the following:

- Run number (sequenced in chronological order)
- Run time: Time from the end of the ripening period to the start of backwash of the filter
- Date and time of the filter run
- Day of the week for the respective filter run
- Number of major and minor turbidity events
- Number of major and minor particle counter events
- Number of spikes in the filter influent
- Significant changes in baseline noise
- Baseline trends
- Peak turbidity value during backwash measured on the influent of the filter
- Measured turbidity value at backwash on the filter effluent stream

Application Note 120

Using this master matrix we were able to quickly identify filter runs that contained events and compare them to filter runs that contained no particle events. From the perspective of this study, an ideal run is one that is free of particle events, trends, or significant baseline noise during the period of time starting at the end of the ripening period and ending at the start of the backwash cycle. In short, the run is stable. Events due to the backwash cycle or within the ripening period were not considered part of the filter run and are not included in the data. Figure 2 shows a "good" filter run, typical of the runs observed over the duration of this study. During this study, 66 percent of the runs were deemed good and had no particle events or trends over these runs.

Filter run termination is primarily determined by time at this water treatment plant. Termination occurs on a regular timed schedule if no breakthrough or loss of head occurs during the prior 20 to 26 hour time frame.

Figure 2 Water Treatment Plant Filter #12 Effluent Particulate Monitoring (05/22/00)

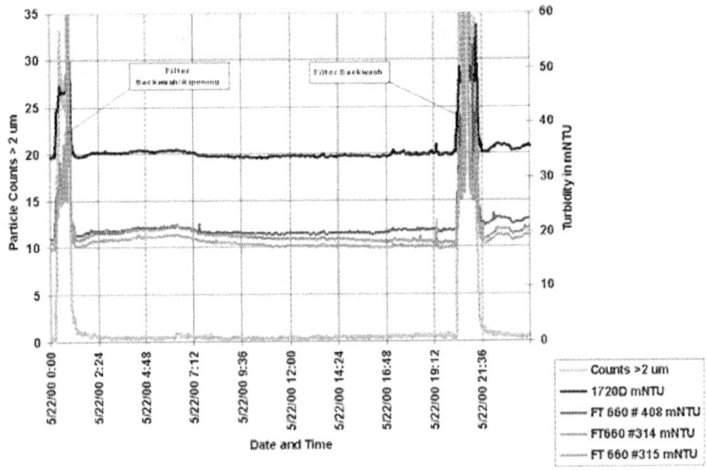

Application Note 120

Of the 66 filter runs, a total of 22 showed at least one particle event. These 22 runs were tabulated into a scaled-down matrix that is summarized in Table 3. The runs are ranked according to the number of total events detected starting with the highest number of events. Within this table, five of the runs had a precursor event that was detected in settled water prior to filtration. An asterisk identifies those runs.

Table 3 Filter Runs showing at Least One Particle Event

EFFLUENT										
			Major Filter Events			Minor Filter Events				
Run	Run Length (hr.)	Total Number of Events	Particle Counter	FT660 #315	1720D	Particle Counter	FT660 #315	1720D	Event Date(s)	Backwash Termination Turbidity (NTU)
21*	24.98	3	3	1	1		2	2	6/1/00	1.74219
54*	24.30	3	—	1	1	2	2	2	7/5 & 7/6	1.10880
58	19.29	3	1	—	—	2	3	3	7/9/00	000.000
2*	26.8	2	2	2	2	—	—	—	5/8/00	0.99504
5	27.05	2	2	—	—	—	2	2	5/17/00	1.48070
22*	31.55	2	2	—	—	—	2	2	6/2/00	0.81269
49	23.95	2	—	—	—	2	2	2	7/1/00	1.05519
52	20.72	2	1	—	—	1	2	2	7/4/00	1.13972
1	20.33	1	—	—	—	1	1	1	5/7/00	1.11836
4	21.23	1	—	—	—	1	1	1	5/16/00	1.17717
12	17.28	1	—	—	—	1	1	1	5/23/00	1.54982
14	22.04	1	1	—	—	—	1	1	5/25/00	1.05917
15*	26.33	1	1	—	—	—	1	1	5/26/00	1.07300
25	22.87	1	—	—	—	1	1	1	6/5/00	1.17753
32	23.19	1	—	—	—	1	1	1	6/13/00	0.94590
34	26.04	1	1	—	—	—	1	1	6/15/00	0.93263
36	27.63	1	—	—	—	1	1	1	6/17/00	0.78922
38	24.79	1	1	—	—	—	1	1	6/19/00	0.69477
44	24.94	1	1	—	—	—	1	1	6/25/00	0.85205
46	35.52	1	—	—	—	1	1	1	6/28/00	0.87059
59	16.53	1	—	—	—	1	1	1	7/10/00	1.05735
63	20.53	1	1	1	1	—	—	—	7/13/00	1.10095

* Runs with an event in the filter influent and in the filter effluent that were detected by the nephelometric turbidimeters 100% of the time and by the particle counter 60% of the time.

Application Note 120

Figure 3 and 4 show filter runs that have either particle events or excessive noise (when compared to the criteria of a good filter run). Figure 3 shows that the events (at 6:30, 8:40, and 9:45) are detected by each of the instruments on the effluent stream. However, the last event before backwash (at 9:45) is detected earlier on the turbidimeters than on the particle counter. This indicates that the sub-micron particles (detected by the turbidimeters) are precursors to larger particles (detected later by the particle counter).

Figure 3 Water Treatment Plant Filter #12 Effluent Particulate Monitoring (05/31/00-06/1/00)

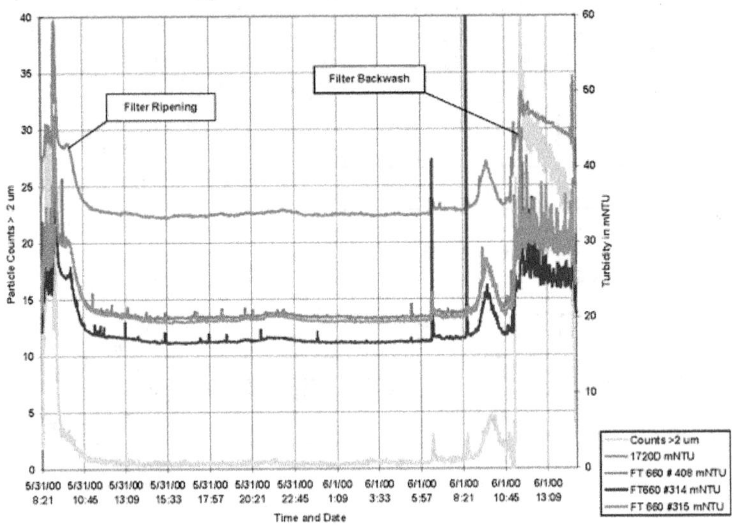

Application Note 120

Figure 4 does not distinguish separate events, but the baseline noise is substantial throughout the run and the particle count trends do not follow the turbidity trends. When comparing all the runs in the study, this run stands out due to the high level of noise and lack of complementary data on the instruments. Reviewing the log books may lead the operator to the cause of the noise.

Figure 4 Water Treatment Plant Filter #12 Effluent Particulate Monitoring (06/20/00)

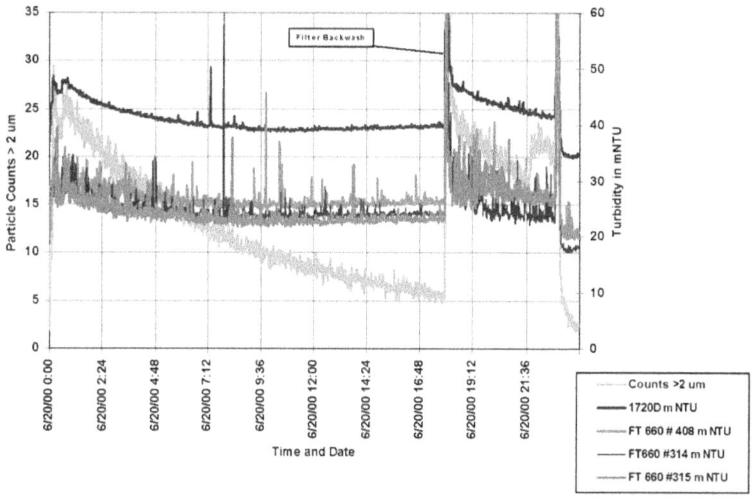

Of the 22 runs that contained particle events, five of the runs contained influent spikes that could be detected in the effluent as well. In all five cases where the spike is detected above the filter, a similar spike in the effluent is seen within the next couple of minutes. Figure 5 shows that the particle event in the influent appears to be a precursor to the particle event that is immediately observed in the effluent with the turbidity technologies. During this run, the spike (at 13:50) in the settled water immediately above the filter increased from 1.2 to 1.9 NTU, a 0.7 NTU increase. The effluent event increased approximately 0.02 NTU, indicating that the filter did remove the majority of this spike. In all five cases, particle spikes that were observed above the filter were easily detected by the laser nephelometer.

Since both the OptiQuant SST and the FilterTrak 660 Nephelometer are calibrated using formazin, the light source differences between the two instruments are minimized. Positioning the instruments on both the influent and the effluent sides of the filter allows log removal calculations to be performed based on the turbidity differential across the filter.

Application Note 120

Color interferences in the influent are eliminated by the 860-nm wavelength of the OptiQuant SST Probe Turbidimeter. Color is not an interference in the low turbidity levels of the effluent stream.

Several spikes that were recorded at the settled water basin by the 1720C turbidimeter were not detected by the instrumentation downstream from this sample point. Hydraulic surges are the suspected cause and these events are short-lived. The particle spikes that were investigated in this study were those that are tracked through the filter into the effluent.

Figure 5 Water Treatment Plant Effluent Particulate Monitoring and Settled Water Applied to Filter #12 (07/05/00)

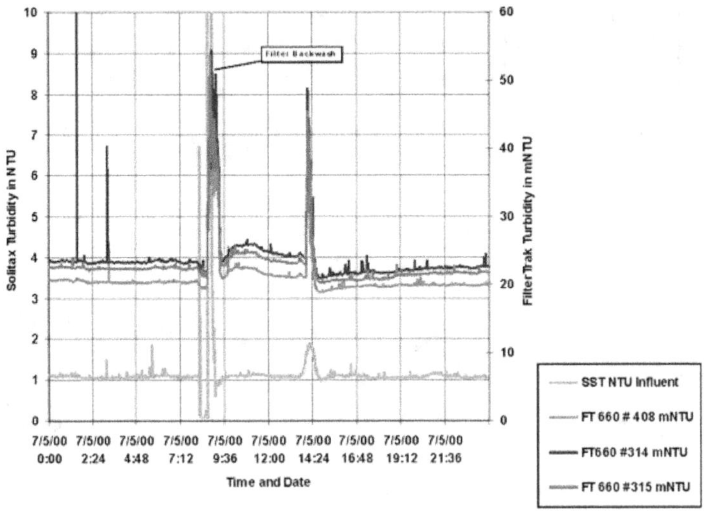

In the majority of the 66 filter runs in this study, an additional occurrence was seen. At the beginning of a typical good run, the FilterTrak 660 baseline showed very low noise. The low noise is maintained until the run is between 65 and 75 percent complete. The noise appeared to increase dramatically as the run progresses toward termination. Figure 6 shows a typical run in which the three FilterTrak 660 Nephelometers all display the same magnitude of baseline noise throughout the run. For 10 randomly selected "good" filter runs (defined as runs without spikes), we looked at the relative standard deviation for the first 75 percent of the run compared to the last 25 percent of the run up to backwash. The baseline relative standard deviation for the last 25 percent of a filter run increased 2.35 times the baseline relative standard deviation over the first 75 percent of the same run.

Application Note 120

It is speculated that large particle detachment from the filter media may be the cause of the increase in background noise as the filter run progresses. If this is true, then monitoring background noise may be another means of predicting breakthrough.

When looking at Figure 6 it is interesting to note that there is one particle event that is detected by the turbidimeters, but is missed by the particle counter (at 15:00). This indicates that the event is primarily sub-micron and is below the detection threshold detection limit of the particle counter.

In the large majority of the filter runs logged, the two technologies—turbidity and particle counting—complemented each other when detecting events in the filter effluent. However, in a couple of cases, events that were detected by the turbidimeters were totally missed by the particle counter. Figure 7 is the same filter run displayed in Figure 5 but includes the data from the particle counter. This figure shows that the precursor event detected above the filter was seen by the effluent turbidimeters, but was not detected by the effluent particle counter. Because the particle counter missed the event, we can surmise that this event is sub-micron in nature.

Figure 6 **Water Treatment Plant Filter #12 Effluent Particulate Monitoring (06/21/00–6/22/00)**

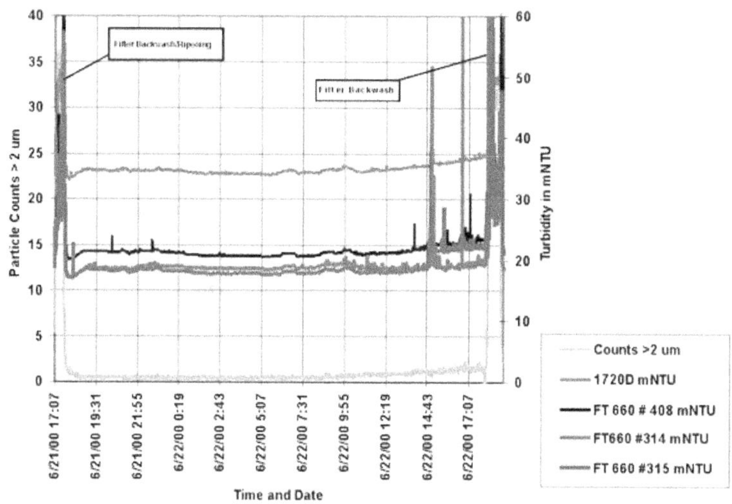

Application Note 120

Figure 7 Water Treatment Plant Filter #12 Effluent Particulate Monitoring (07/05/00)

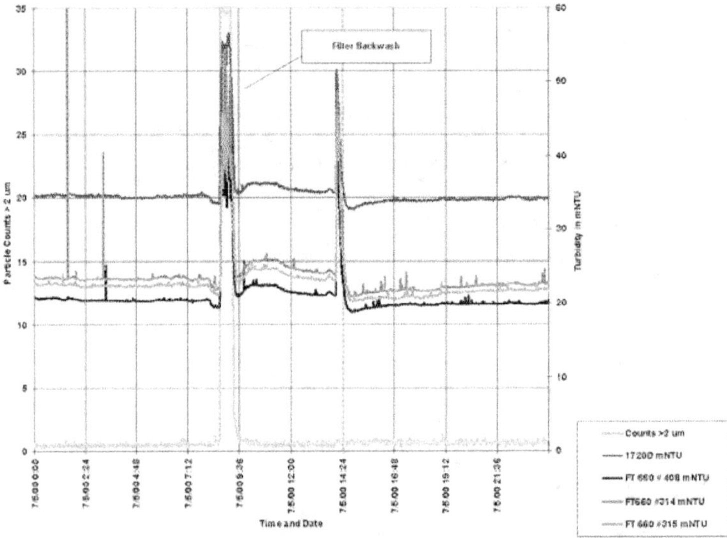

Correlations between the turbidity value at the termination of backwash and filter events were also examined. The turbidity values at backwash termination were in a very narrow range between 0.69 and 1.74 NTU. It can be speculated that the tight range of values at backwash termination predicts the overall consistency of the filter runs over time. No correlation was found between these values and particle events. The correlation between the peak turbidity at backwash and subsequent filter events of the proceeding run was also investigated. Again, there was no correlation between these two parameters.

Conclusions

Of the 66 continuous filter runs that comprise this study, 33 percent contained particle events as they were defined at the beginning of this study. Of the 22 runs that contained events, 23 percent appear to have a precursor event detected by the OptiQuant SST probe turbidimeter that was monitoring the pre-filter sample.

The FilterTrak 660 detected the majority of events that the other effluent instruments detected and also detected some events that they missed. In addition, the FilterTrak 660 baseline standard deviation increases as the run progresses. This may be a precursor to breakthrough and warrants further investigation.

Application Note 120

Events that are detected in the pre-filtered water were also consistently seen by the FilterTrak 660 and particle counter. All five settled water precursor events were also seen in the effluent. The turbidity of these influent spikes, which range between 0.5 and 2 NTU, were reduced significantly as they passed through the filter. The resulting events in the effluent were very small with turbidity changes ranging between 0.005 and 0.030 NTU (5-30 mNTU) and the finished water was maintained far below the requirements of the Partnership for Safe Drinking Water.

The impact on construction and the seasonal drought in the area did not appear to correlate to the frequency of events. Runs with events did occur in an apparently random order during the 66 runs.

As was discussed earlier, having both a particle counter and a laser nephelometer provides information as to the composition of a particle event. Event detection that is complemented by both instruments indicates a natural distribution of particles roughly following the $1/d^3$ relationship. In these cases, the nephelometer will detect the particles slightly before the particle counter because small particles move more rapidly through a filter. If only the laser nephelometer detects the event, the composition of the particles is most likely sub-micron in nature. If the particle counter alone detects the event, this indicates a non-natural distribution of large (2 µm) particles and may indicate a change in the conditions within the filter or a contamination issue. In all cases, the use of two instruments provides further insight into the particle sizes of respective events.

This WTP filter effluent did not exceed the Partnership turbidity limits throughout the entire study (including backwash runs). However, the instrumentation did show both good, clean filter runs along with runs with definitive particle events. Though it may be challenging, the WTP management can investigate their logs to see if the runs that contained events relate to any changes in the treatment upstream of the filter. This is continuous improvement at its best.

The WTP showcased in this study is, in reality, a best-case scenario. Its processes were optimized for the duration of this study and are under very tight control at all times. However, a WTP that does not have consistent filter runs, or one that often has particle spikes could use this instrumentation to detect, analyze and eventually reduce or eliminate such events. The intent and anticipated use of the study instrumentation goes beyond regulatory requirements and will help plants achieve production of water characterized by high quality and consistency.

We plan to continue monitoring this filter for the benefit of the plant management. Due to structural problems on the dams for the raw water source, the source will be drained significantly throughout the summer and fall of 2000. We will continue to see if changes to the raw water source have an impact on particle event occurrence at this sample site.

APPENDIX F

Verifying the Calibration of Particle Counters

Michael Sadar, Research Scientist, Hach Company

Particle counters are fast becoming an important process monitoring tool to verify the quality of effluent water as it leaves the filter. Because these instruments are somewhat complex, it is important for operators to become familiar with the theory behind their operation and calibration. Up to now, particle counter calibration has not typically been performed by the user, due to the expensive apparatus and technical expertise required.

The Importance of Calibration

While it may not be necessary to calibrate at routine intervals, it is imperative that the instrument's calibration be verified. Many of the available methods do not necessarily test the validity of the particle counter calibration. To do so, the following criteria must be met:

- The verification must use the same standard material used in the calibration. Most particle counters are calibrated with polystyrenelatex (PSL) spheres. These spheres have a defined size, shape and refractive index. Verification with a material that has a different size, shape and/or refractive index will introduce significant error.
- The statistics used in calibration should also be applied in verification. When particles in a standard are sized during an instrument calibration, this sizing accuracy is assumed to fit a Gaussian model. (In reality, the count distribution is not exactly Gaussian.) The application of a Gaussian model allows for several tests to be developed that can verify whether or not the instrument meets its performance criteria.
- In a Gaussian model, it is given that a mono-dispersed standard be analyzed. Such a standard must have a very narrow distribution of a specific size. When a mono-dispersed standard is analyzed by a particle counter, Gaussian statistics allow us to postulate that 67 percent of all the analyzed standard will be sized to within one standard deviation from the mean size of the standard, and that 95 percent of all particles analyzed will be sized to within two standard deviations from the mean size of the standard.

How Is the Standard Deviation Defined?

The standard deviation is a variable in the equation that defines instrument resolution:

Resolution (R) is defined as one standard deviation (σ) divided by the size of the particle (m), or

$$R = \sigma / m \qquad (1)$$

Thus, if resolution is defined (or if the resolution is to be tested for) and the size of the particles in the standard is known, then the standard deviation can be calculated by rearranging equation 1:

$$\sigma = Rm \qquad (2)$$

Once the standard deviations are defined, size bins can be set to encompass ±1 σ and ±2 σ from the mean size of the standard used. Thus, a verification test can be designed with three specific size bins, which will be used to verify both count accuracy and size accuracy. One bin counts particles that are ±1 standard deviation from the mean; another bin counts particles from one to two standard deviations above the mean; and the third bin counts particles from one to two standard deviations below the mean.

Count Accuracy Verification:

The 1900 WPC bin sizes are set at one and two standard deviations from the mean size of the standard injected during verification. The mean diameter of this standard is 10.0 µm ±0.09 µm, which allows it to used as a mono-dispersed size distribution. The three bins are then defined at 7.0 to 8.5 µm, 8.5 to 11.5 µm, and 11.5 to 13 µm, which collectively encompass ±2 standard deviations from the mean. This set-up will test for 15 percent resolution at 10 µm*.

Statistically, 95 percent of all injected particles should fall within two standard deviations about the mean. Thus, if 2000 counts are injected, at least 1900 counts should be detected in the three size bins that cover the size range of 7 to 13 µm (plus or minus test errors associated with the standard, injection, and instrument).

Size Accuracy Verification:

The same three bin sizes are used to define size accuracy (resolution). Bins are set up according to resolution with three bins representing two standard deviations about the mean. Using Gaussian statistics, 67 percent of all particles counted must fall within the first standard deviation about the mean particle size of the standard. When errors for the verification procedure are taken into consideration, the pass/fail condition for size resolution is adjusted

*1900 WPC Sensor has <10% resolution at 10 µm. The 15% resolution test was designed to determine a deviances of >5% from this specification.

to 60 percent. If more particles fall within this first standard deviation bin, the resolution is better than 15 percent (pass)'. If fewer particles fall into this first standard deviation bin, then the resolution is poorer than 15 percent (fail).

To calculate whether or not the instrument passes the size accuracy (resolution) specification, divide the number of counts in the central bin (± one standard deviation from the mean) by the total number of counts that fall within ± two standard deviations from the mean. A number of 0.6 or greater indicates a resolution equal to or better than 15 percent, and the instrument passes.

New Standard Additions Procedure

The 1900 WPC calibration verification procedure is based on the method of standard additions. In this method, a defined volume of a standard is injected into a particle counter and analyzed. The standard itself is defined with respect to size and counts of the particles. Thus, if a known number of particles are injected and the size of these particles is also known, both count and size accuracy can be tested using the above Gaussian model. The following sections describe the setup of a particle counter for such a calibration verification.

The 1900 WPC calibration verification procedure tests for both count and size accuracy (resolution). There are several errors associated with the verification test. These include the standard (±10 percent on count), injection accuracy (±5 percent), the instrument (±10 percent on size and counts), and baseline variability. Statistical propagation results in an excess of 20 percent with respect to counts. Thus, the established pass/fail conditions for verification have been set at ±20 percent, which constitutes a range of 1520-2280 counts for a pass.

What if the Instrument Passes?

Under a pass condition, the instrument will pass size accuracy to better than 15 percent at 10 µm. If desired, a more strict test could be performed to test for either tighter resolution or to test resolution using a different size standard. In either case, the standard deviations about the mean value of the new standard would have to be calculated and size bins be re-set accordingly.

What if the Instrument Fails?

There are several causes for instrument failure. The most common causes are low count and/or under-sizing failures and high count failures. These are discussed below:

The most common symptom of verification failure is the combination of under-sizing and under-counting of the standard. Counts will fall below the pass/fail range and the particles in the centralized bin will add up to less than 60 percent of the total. It is common to observe more counts in the -2 to -1 standard deviation bin (7.0 to 8.5 µm) than the central bin. The most common cause of this symptom is a dirty sample cell. Cleaning the sample cell usually alleviates the condition. A second cause of under-counting symptom is an excessive flow rate. Though not common, other sources of failure related to these symptoms can be attributed to a damaged sample cell, misaligned optics, or a decrease in laser output from its source.

The best approach to eliminate sources of failure is to thoroughly clean the instrument, specifically the sample cell portion of the sensor. It is also important to verify that the flow rate is correct (200 ±10 mL/minute). Then repeat the verification procedure. If the instrument fails again, the problem may be more serious. A service person should be contacted or the instrument may need a factory calibration (Hach recommends sensor calibration every 12 months).

Failures with the symptom of over-counting can also occur. Likely causes are contamination of the standard, injection apparatus, or particle shedding in the inlet sample lines during verification. Another cause could be a low-flow condition. There is a small probability that the loss of laser output could result in an over-counting symptom.

The first approach to eliminating the high-count symptom is to first verify the flow rate. Next, repeat the verification using a new verification kit. Care must be taken to prevent contamination of the standard, apparatus, or instrument. A second failure may indicate a more serious condition. A service person should be contacted or the instrument may need a factory calibration.

Summary:

Particle counter calibration verification can be used to provide confirmation that the operational state of the tested instrument is correct and acceptable. Further examination of a failure can often be corrected at a minimal cost of time and labor to the operator. Verification is a valuable and effective tool that can be used to ensure the particle counters are performing within their specifications.

APPENDIX G

Table G-1 Unit conversion factors, SI units to US customary units, and US customary units to SI units

SI unit name	To convert, multiply in direction shown by arrows				US customary unit name
	Symbol	→	←	Symbol	
		Volume			
Cubic meter	m^3	1.3079	0.7646	yd^3	Cubic yard
Cubic meter	m^3	264.1720	3.7854×10^{-3}	gal	Gallon
Cubic meter	m^3	8.1071×10^{-4}	1.2335×10^{3}	acre-ft	Acre-foot
Liter	L	0.2642	3.7854	gal	Gallon
Liter	L	0.0353	28.3168	ft^3	Cubic foot
Liter	L	33.8150	2.9573×10^{-2}	oz	Ounce (U.S. fluid)

NOTE: cm is not a SI unit but is included because of its common usage.

UV Dose Conversion

1 mW-sec/cm^2 = 1 m^3/cm^2 = 10 J/m^2

Table G-2 Common water treatment conversion factors, SI units to US customary units, and US customary units to SI units

To convert, multiply in direction shown by arrows					
SI unit name	**Symbol**	→	←	**Symbol**	**US customary unit name**
		Concentration			
Kilogram per cubic meter	kg/m^3	8.34×10^3	1.2×10^{-4}	lb/Mgal	Pounds per million gallons
Milligram per liter	mg/L	8.34	0.12	lb/Mgal	Pounds per million gallons
		Flow rate			
Cubic meters per day	m^3/d	2.642×10^{-4}	3.785×10^3	mgd	Million gallons per day
Megaliters per day	ML/d	0.2642	3.785	mgd	Million gallons per day
Liters per second	L/s	15.852	0.0631	gal/min	Gallons per minute
		Hydraulic loading rate			
Cubic meter per square meter-hour	m^3/m^2-h	0.4098	2.44	$gal/min/ft^2$	Gallons per minute per square foot
Meter per hour	m/h	0.4098	2.44	$gal/min/ft^2$	
		Mass loading			
Kilogram per day	Kg/d	2.2046	0.45359	lb/d	Pound per day
		Pressure (force/area)			
Kilopascal	kPa	0.1450	6.8948	lb/in^2	Pounds per square inch (psi)
Bar	bar	14.504	0.06895	lb/in^2	Pounds per square inch (psi)

INDEX

Note: *f.* indicates figure; *t.* indicates table.

www.ingramcontent.com/pod-product-compliance
Lightning Source LLC
LaVergne TN
LVHW052327200926
848481LV00002B/689